Building Pathology and Rehabilitation

Volume 25

This book series addresses the areas of building pathologies and rehabilitation of the constructed heritage, strategies, diagnostic and design methodologies, the appropriately of existing regulations for rehabilitation, energy efficiency, adaptive rehabilitation, rehabilitation technologies and analysis of case studies. The topics of Building Pathology and Rehabilitation include but are not limited to - hygrothermal behaviour - structural pathologies (e.g. stone, wood, mortar, concrete, etc...) - diagnostic techniques - costs of pathology - responsibilities, guarantees and insurance - analysis of case studies - construction code - rehabilitation technologies - architecture and rehabilitation project - materials and their suitability - building performance simulation and energy efficiency - durability and service life.

J. M. P. Q. Delgado
Editor

Building Pathologies: Experimental Campaigns and Numerical Procedures

Springer

Editor
J. M. P. Q. Delgado
CONSTRUCT-LFC
Department of Civil Engineering
Faculty of Engineering
University of Porto
Porto, Portugal

ISSN 2194-9832 ISSN 2194-9840 (electronic)
Building Pathology and Rehabilitation
ISBN 978-3-031-17063-8 ISBN 978-3-031-17061-4 (eBook)
https://doi.org/10.1007/978-3-031-17061-4

This Springer imprint is published by the registered company Springer Nature Switzerland AG
The registered company address is: Gewerbestrasse 11, 6330 Cham, Switzerland

Preface

In the distant past, construction solutions were validated empirically through years of experience; however, it was equally apparent to those promoting such novel approaches that selection based on an understanding of the performance requirements could only be met if the results of research and development were made available and indeed exploitable by practitioners. To provide an economic and effective remedy for building defects, it is essential to properly identify the cause in order to address the problem. Rehabilitation is a strategic area that is concerned not only with historic buildings but also with other buildings that have been in use for some time and need to be adapted to the demands of the present.

The main purpose of this book, *Building Pathologies: Experimental Campaigns and Numerical Procedures*, is to provide a collection of recent research works related to building pathologies, in order to contribute to the systematization and dissemination of knowledge related to construction pathology, hygrothermal behavior of buildings, durability, and diagnostic techniques and, simultaneously, show the most recent advances in this domain.

The book is divided into six chapters that intend to be a resume of the current state of knowledge for benefit of professional colleagues, scientists, students, practitioners, lecturers, and other interested parties to network. At the same time, these topics will be going to the encounter of a variety of scientific and engineering disciplines, such as civil, mechanical, and materials engineering.

Porto, Portugal

J. M. P. Q. Delgado

Contents

Numerical Simulation of Moisture Transport Along Ceramic Bricks—Wetting Process

C. M. Araújo, A. C. Azevedo, and F. A. N. Silva

Abstract The moisture transport in brick masonry is an important phenomenon in several deterioration mechanisms. However, is a very complex process and is influenced by many physical phenomena. Investigation of the moisture transfer through a building wall, which in general, consists of multiple layers, presumes knowledge about the continuity between layers. In this study, three types of contact configurations were analysed, as follows: Hydraulic contact, Perfect contact, and Air space. Therefore, to understand the moisture transport in brick masonry the moisture transport through the interface of materials was analyzed. This was done for samples of brick-cement mortar, brick-lime mortar, and air space between brick-layers, as well as for samples with different interface location heights and different mortar thicknesses and air space. Mainly, the present work is concerned to simulate the hygrothermal behaviour across brick–mortar and brick-brick interfaces to compare the results with the laboratory analysis. The numerical simulations of brick–mortar and brick-brick samples were performed with the hygrothermal simulation software WUFI-2D. The data used to run the simulations were taken from the wetting experiments on the samples; the corresponding moisture content profiles were measured using gamma-ray spectrometer. Although the mechanisms of moisture transport in a single building material have been and continue to be extensively studied, the hydraulic characteristics of the interface at different types of contacts between materials are still poorly comprehended and for this reason, the simplified assumption of perfect contact, is widely used in hygrothermal models. In general terms, the assumption of perfect contact implies that the interface will have no effect on moisture transport. In comparison, the imperfect contact assumption implies that the interface between building materials will resist moisture transport. However, comparisons between experimental

C. M. Araújo · F. A. N. Silva
Civil and Engineering Department, Catholic University of Pernambuco, Recife, Pernambuco, Brazil
e-mail: camila.2020604019@unicap.br

F. A. N. Silva
e-mail: fernando.nogueira@unicap.br

A. C. Azevedo (✉)
Instituto Federal de Ciências de Educação e Tecnologia de Pernambuco (IFPE), Recife, Brazil
e-mail: antonio.costaazevedo@fe.up.pt

J. M. P. Q. Delgado (ed.), *Building Pathologies: Experimental Campaigns and Numerical Procedures*, Building Pathology and Rehabilitation 25,
https://doi.org/10.1007/978-3-031-17061-4_1

and numerical results were used to investigate whether the perfect contact assumption is appropriate for real samples.

Keywords Ceramic brick · Numerical simulations · WUFI · Moisture transport · Interface

1 Literature Review

1.1 Moisture Transfer Concepts

1.1.1 Material Properties for the Moisture Transfer Study

The envelope of any building is continuously exposed to changes in internal and external temperature, pressure, and humidity conditions. This results in an exchange of energy and mass (air and moisture) between the interior and outdoor environments through the envelope. Experts in the literature refer to this phenomenon as "heat, air, and moisture transport" through building materials and components.

For the advancement in building construction, considering a reduction in generated energy consumption and increased durability of components, designers and builders are always interested in knowing the long-term performance of the building envelope as subjected to transport processes.

However the global differences in building practices, construction materials, climatic conditions, and indoor climate are so large, it is impractical to develop this knowledge only through experimental investigations.

Therefore, nowadays the knowledge of the performance of building components, as well as the whole building, is already possible due to the advancement of technology and is being improved over time through the development of calculation methods for this purpose.

However, the diverse set of procedures and computational models require information regarding the materials that are constantly evolving. The main consequence is a remarkable change in the properties of these materials. Therefore, the properties reported in the literature may become "unrepresentative" of actual products. This requires a continuous update of the hygrothermal properties. Otherwise, however, the sophisticated numerical calculation method employed may produce results that do not represent the real hygrothermal behavior of the building component under analysis.

The main hygrothermal properties of materials, considered in most of the existing numerical models for the analysis and knowledge of the building components' performance undergoing moisture transport, will be presented below according to Kumaran (1996) review.

- *Density* (ρ_0—kg/m^3)

The density of given building material is obtained by the ratio between the mass of a dry specimen and its total volume. The drying of the specimens should be done in an oven until the material reaches a constant mass. For the determination of the volume, it is necessary to measure the dimensions of the specimen, performing three readings per measurement and calculating its arithmetic mean. The volume is calculated by multiplying the three dimensions of the specimen (corresponding to three arithmetic means).

- *Moisture Content* (w)

In building materials with a porous structure, moisture transfer takes place through the networks and capillary channels of its constituents. As a rule, these materials are hygroscopic, i.e., when materials are placed in an environment where the relative humidity varies, their moisture content also varies.

In the presence of air humidity, these materials tend to retain water vapor by adsorption until they reach a given state of equilibrium with the environment—hygroscopic equilibrium—the respective moisture content being called hygroscopic moisture content. This phenomenon is attributed to intermolecular or Van der Waals forces acting on the solid–liquid interface, inside pores, as shown in the hygroscopic curve (see Fig. 1).

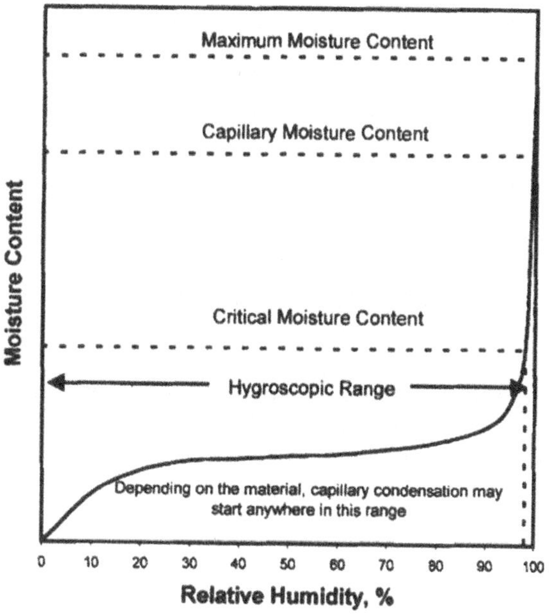

Fig. 1 Sorption Isotherm of porous building material

The hygroscopic curve has three phases. In a first phase, a layer of water molecules is fixed on the inner surface of the pores—monomolecular adsorption—followed, in a second phase, by the deposition of several layers of molecules—multi-molecular adsorption. When the diameter of the pores is small enough, the multi-molecular layers join together—capillary condensation.

When the environment in contact with the element is saturated, the hygroscopic equilibrium moisture content is called the critical moisture content (w_{cr}). If it is put in contact with liquid water, its moisture content increases (by filling the pores and channels with liquid water) and this filling is carried out by capillarity until reaching the saturation value, which corresponds to the maximum moisture content, w_{sat}. Above this value, it is only possible to increase its water content as a result of external pressures. The maximum moisture content is hardly reached given the ease of air being trapped inside the porous structure. For this reason, the capillary saturation moisture content, wcap, is defined, which represents the maximum amount of moisture that a material contains after contact with liquid water.

The moisture content of a material, in contact with air, can thus vary between absolute zero (if the environment has zero relative humidity) and the critical moisture content.

- *Specific heat capacity* (c_0—J/(kg K))

The specific heat capacity is the amount of heat needed to change the temperature of the unit of mass of a given material. The method typically used for the determination of the specific heat is called the Method of Mixtures. This is based on the "principle of equality of heat exchanges", which dictates that: when heat exchanges between bodies that are thermally insulated from the outside environment, the amount of heat released by bodies that cool is equal to the amount of heat received by bodies that heat up.

The test uses water as the calorimetric fluid at an average temperature of 60 °C (range 20–100 °C)—the use of other fluids can vary the range, depending on what is intended. The specimens are subjected to drying in an oven at a temperature that varies between 102 and 120 °C. Subsequently, a given mass of material is added at elevated temperature, to a given mass of low-temperature water (mixing method). Therefore, the equilibrium temperature is determined resulting. Calculating the heat absorbed by the water and the container and equating its value to the expression of the amount of heat released, we were able to determine the value of the specific heat capacity. Specific heat capacity is designated by the symbol c0 and is usually expressed in the units J/(kg K). If the material is wet, the specific heat capacity c is to be calculated as:

$$c = c_0 + 4187 \cdot \frac{W}{\rho_0} \tag{1}$$

The above relation assumes that the specific heat capacity of water is a constant equal to 4187 J/kg K.

- *Thermal conductivity* (λ—W/(m K))

Thermal conductivity is the heat flux (in Watts) that crosses, in a perpendicular way, an element with a section of 1 m^2 and a thickness of 1 m, when the temperature difference between the two faces is 1 °C, which are flat and parallel to each other. The definition of thermal conductivity from Fourier heat conduction is given by:

$$q = -\lambda \cdot \text{grad}\,T. \tag{2}$$

The Guarded-Hot-Plate Method is one of the most used for the determination of thermal conductivity. Its procedure involves, firstly, placing two equal specimens with smooth faces (30 × 30 cm^2 and 7 cm maximum thickness) between metal plates, placing an insulat on the side to prevent the flow of heat through that side. After a period of stabilization of the system, a constant and unidirectional flow is obtained, in the direction perpendicular to the faces of the specimen. The thermal conductivity value is determined by the knowledge of the geometry of the specimens, the values of the electrical power supplied to the main heater and the values of the temperatures on the faces of the specimens.

- *Water Vapor Permeability* (δ_p—kg/msPa)

The water vapor permeability is the amount of vapor that passes through a unit of time, in a steady state, through the unit surface of the material with unit thickness, when the vapor pressure difference between the two parallel faces is also unitary.

There are different experimental methods for the determination of this property, but the most traditional methods are the dry cup method, and the wet cup method. A test vat impermeable to water vapor, sealed on the open side, containing a test piece and a saturated saline solution—Wet Cup Method—or, alternatively, a desiccant—Dry Cup Method, is placed in a climatic chamber.

A relative humidity difference is imposed in a climatic chamber and, due to the partial vapor pressure gradient established between the sample and the chamber, a vapor flow is developed through the specimen. The steam flow is determined through the specimen by successive weighing's. Thus, it is usual to distinguish the two methods according to the following differences in relative humidity between the faces of the specimen:

- Dry Cup method—for relative humidity between 0 and 50%;
- Wet Cup Method—for relative humidity between 50 and 100%.

Several research studies used the vapor resistance factor of a material. This variable is defined as the ratio between the water vapor permeability of the material and the vapor permeability of the stagnant air, δ_a, as:

$$\mu = \frac{d_p}{\delta_a} \tag{3}$$

The vapor permeability of stagnant air can be calculated according to the given by Schirmer (1938):

$$\delta_a = \frac{2.306 \times 10^{-5} P_0}{R_v T P}\left(\frac{T}{273.15}\right)^{1.81} \tag{4}$$

where, T is the temperature (K), P is the ambient pressure (Pa), P_0 is the standard atmospheric pressure (i.e., 101–325 Pa) and R_v is the ideal gas constant for water [i.e., 461.5 J/(K kg)].

- *Porosity* (ε—m^3/m^3 or %)

The porosity is a physical property of the material and it is defined by the ratio of the volume of the material's pore to its total volume. This dimensional variable it is very useful to determine the maximum water content w_{max}, through the multiplication of the porosity value by $\rho_w = 1000\ kg/m^3$.

- *Water absorption coefficient* (A_w—$kg/(m^2\sqrt{s})$)

The water absorption coefficient characterizes the capillarity of materials, that is, the property of a material to absorb liquid water by suction when it is in contact with water.

The laboratory methodology for determining this property provides that, firstly, the specimens are placed in the environment where the tests will be carried out, at a certain relative humidity, until their mass stabilizes. That said, the specimens must be waterproofed on their side faces, in order to obtain a unidirectional flow. Then they must be dipped in a vat with water, trying to keep the bases immersed in water (52 mm). In the first 24 h, the weighing's are made periodically following a logarithmic scale. After this time, the weighing's are carried out for periods of 24 h. The absorption coefficient is determined from the linear relationship between the square root of time and the variation in the weight of the specimens. If there is no linear relationship, the determination is carried out with values recorded at 24 h.

1.1.2 Transport Mechanisms

The transport of heat, air, and moisture have a complex physics involved. Heat and mass transport occurs simultaneously in each porous construction material. The interaction of one or more moisture phases: vapor, liquid, and solid ice, if present, can interact with the porous media's solid matrix phase. Besides the phase change physical phenomena such as evaporation, condensation, absorption of heat, freezing and thawing, can also occur during moisture transport.

This section aims to provide a theoretical review of the transport mechanisms present in hydrothermal processes in porous media. Several authors (Kuenzel 1995, 1996; Kohonen 1984; Janssens 1998; Luikov 1966) provide in their work more details on the theoretical development of various transport potentials. The choice of moisture transport potential was made based on what was familiar to the author of the review.

Many older models of moisture transport were developed based on discontinuous potentials, such as moisture content (Luikov 1966; Pel 1995; Trechsel 2001) reached the conclusion that the moisture content at an interface between two materials is discontinuous. Therefore, a more extensive analysis at the inhomogeneous material intersection is required.

- *Mass Transfer*

Kerestecioglu et al. (1989) and Kaviany (1993) provide a great description of the transport mechanism in porous media. However, it must be considered that the transport coefficients may not only be strong functions of the independent variable, but may change as a function of time and exposure. Some of the most important ways in which moisture can be transported are:

- *Molecular vapor diffusion*, by partial vapor pressure gradients.
- *Molecular liquid diffusion*, movement of the liquid phase due to liquid-filled pores.
- *Capillary liquid flow*, movement of the liquid phase due to capillary suction pressures.
- *Knudsen vapor diffusion*, movement of the vapor phase in small pores and at low pressures; the mean free path is greater than the pore diameter and collisions of molecules with the pore walls occur more frequently than collisions with other diffusing molecules.
- *Evaporation–condensation vapor flow*, movement occurs in conjunction with heat transfer, moisture evaporates and re-condenses in a similar fashion to a heat pipe.
- *Gravity-assisted diffusion liquid flow*, movement occurs due to gravity and occurs mostly in macro-porous materials.

- Vapor Transport

The diffusion of water vapor under isothermal conditions may be described by Fick's first law for unimpeded flow in still air (Belarbi et al. 2006):

$$q_v = -D_v \nabla X \tag{5}$$

where q_v is the mass flux rate of vapor flow (kg/m^2 s), D_v the diffusion coefficient of vapor in air m^2/s, and X the vapor concentration (kg/m^3).

$$q_v = -D_v \frac{M}{RT} \nabla P_v \tag{6}$$

where R is the universal gas constant (8.314 J/molK), P_v is the partial vapor pressure, T is the temperature (K), and M is the molar weight of water (0.018 kg/mol).

In a porous material, diffusion is reduced in comparison to that in still air by a resistance that corresponds to the volume fraction of air-filled open pores a and α

tortuosity factor α. This is expressed as:

$$q_v = -\alpha \, a \, D_v \frac{M}{RT} \nabla P_v \tag{7}$$

European countries mostly introduce a resistance factor as $\mu = (1/a \cdot \alpha)$ leading to the following flux for vapor flow:

$$q_v = -\frac{D_v}{\mu} \frac{M}{RT} \nabla P_v \tag{8}$$

- *Liquid Transport*

The liquid flow is transported differently within the two regions of interest in building materials and can be defined in two ways: Capillary water region, which follows the hygroscopic sorption region and extends to free water saturation. This region can be characterized by equilibrium states. Liquid transport occurs under the influence of a pressure or suction force in the capillary regime, and the transport of wet liquids occurs mainly in this region; Supersaturated capillary region, follows the capillary water region. Normal suction processes are not physically plausible in it. Liquid flow in this region occurs by diffusion under a temperature gradient or by external pressure under suction. In this region, there are no equilibrium states (Trechsel 2001).

The laminar transport theory in capillary tubes developed by Darcy (1856) more than 150 years ago is still the most widely considered in formulations to explain liquid transport. Karagiozis (2001) extended the original formulation, to take into account the forces of gravity, thus the liquid transport in the capillary regime is given by:

$$q_w = -\left(D_\Phi \nabla \Phi + \frac{\left(D_\Phi \frac{\partial \Phi}{\partial u} \right)}{\frac{\partial P_h}{\partial u}} \rho_w \vec{g} \right) \tag{9}$$

where q_w is the mass flux of liquid (kg/m^2 s), D_ϕ is the liquid coefficient (kg/m s), g is the gravitational acceleration (m/s^2), and ρw is the density of water (kg/m^3). The suction pressure is usually described by employing a cylinder capillary model and can be presented as):

$$P_h = 2\sigma \frac{\cos\theta}{r} \tag{10}$$

where σ is the surface tension of water, *r* the capillary radius (m), and θ the contact angle or wetting angle (°).

Using thermodynamic equilibrium conditions for a cylinder capillary model, the relationship between relative humidity ϕ over a concavely curved water surface and the capillary pressure P_h is defined by Kelvin's equation as:

Table 1 Energy transport

Conduction	Convection	Radiation
$q_c = -k\nabla T$ where k = thermal conductivity (W/m K) T = Temperature (°C)	$q_a = \nabla \rho_a \upsilon C_a T$ where ρ_a = Density of air (kg/m^3) v = velocity (m/s) C_a = volumetric heat capacity (J/m^3 K)	$q_r = \varepsilon\sigma F(T_s^4 - T_\infty^4)$ where ε = emissivity of gray surface (–) σ = Stefan Boltzmann constant = (5.67 × 10^{-8} W/m^2 K^4) F = view factor (–) T_b = surface temperature (K) T_a = surrounding temperature (K)

$$\Phi = \exp\left(\frac{-P_h}{\rho_w R_v T}\right) \tag{11}$$

where ρw is the density of water (kg/m^3), P_h is the capillary pressure (Pa), R_v is the gas constant for water vapor (J/kg K), temperature T is in Kelvin (K) and ϕ is relative humidity (–).

- *Heat Transfer*

Heat transfer can occur by conduction, convection, and radiation transfer within building envelopes. Table 1 lists the equations of state that govern these modes of heat transfer. The thermal conductivity k is a function of the ice and liquid content present in the porous material and can be strongly influenced by temperature, as can be direction dependent.

- *Phase Change*

The phase change conversion enthalpies contribute a local source of heat that is stored or released in a porous material when moisture accumulation or drying is present. If one considers that the amount of moisture I_{ij} of phase i is converted to j the control volume receives the following amount of heat:

$$q_r = \Delta h_{ij} \cdot I_{ij} \tag{12}$$

where Δh_{ij} is the change in enthalpy, 3.34 × 10^5 J/kg for conversion of ice to water and 2.45 × 10^6 J/kg at 20 °C for water to vapor.

This quantity of heat may be significant when drying or accumulation is present in porous media. A modeling challenge exists to properly accommodate this latent heat term in the governing equation of energy (Trechsel 2001).

- *Air Transport*

Airflow is driven by a difference of air pressure. The mass flux of air through a porous material may he expressed as:

$$m_a = -k_a \nabla P_a \tag{13}$$

where m_a is the air mass flux (kg/m^2s), k_a is the air permeability (kg/msPa), and P_a is the air pressure (Pa).

1.1.3 Moisture Transport in Porous Media

A porous building material can contain a multiphase combination of solids, liquids, and gas. Where heat and mass transfer occurs essentially at the microscopic level for each phase. Therefore, to solve the transport mechanisms at the microscopic level, a detailed description of the geometry and topology of the porous medium is required. This description of the porous medium and the associated transport processes at the microscopic level is difficult and complex, which results in most analyses of a porous material through laboratory characterizations being performed at the macroscopic level.

Employing this basic phase assumption allows the development of differential equilibrium equations for mass, momentum, and energy transfer in numerical models. The porous medium can then be described by three distinct phases, the solid, the liquid water, and the gas phase.

This work has previously described all the transport mechanisms in a porous material. This knowledge is important for understanding moisture transport using tools that simultaneously simulate Heat and Moisture transport in a material. However, since the object of study in this paper is focused on moisture transport, the two main forms of transfer will be further specified below.

- *Moisture sorption isotherms*

Building materials are mostly hygroscopic as they absorb water vapor from the environment until equilibrium conditions are reached. This behavior can be described by sorption curves over a humidity range between 0 and 95% RH. Materials that have water content that is not very sensitive to temperature changes have sorption curves called sorption isotherms. Sorption curves and sorption isotherms for these materials from 95% RH to capillary saturation at 100% RH are difficult to measure. In this range, the equilibrium water content of a material is still a function of relative humidity. It is therefore difficult to determine the function by sorption tests in climatic chambers and makes it necessary to use a pressure plate apparatus in order to complete the sorption curve in the high humidity range.

The resulting water retention curve has great importance in the study of moisture transport through numerical modeling and is a prerequisite for simulations including liquid transport.

The sorption isotherms are the equilibrium moisture content states in a porous material as a function of relative humidity at a particular temperature. Families of sorption isotherms that encompass both the hygroscopic and capillary regimes are:

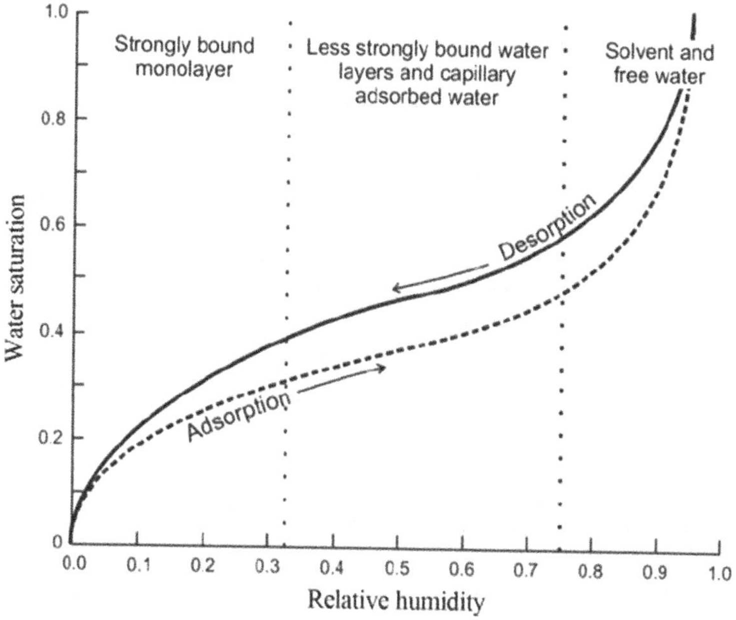

Fig. 2 A schematic representation of a sorption isotherm with a hysteresis between the adsorption and desorption isotherms

- Absorption isotherm and desorption isotherm.
- Hysteresis isotherms (the equilibrium moisture content curves that span the complete spectrum of moisture equilibrium during both absorption or desorption).
- *Temperature-dependent sorption curves* (the equilibrium moisture content curves dependent on temperature).

A schematic representation of a typical sorption isotherm with a hysteresis between the adsorption and desorption isotherms is plotted in Fig. 2.

These two relationships can be transformed from one to the other on the basis of Kelvin's equation. That means there is an intrinsic relationship between capillary pressure, water saturation and relative humidity.

The sorption consists of adsorption and desorption. The adsorption is induced by interaction forces between water molecules and solid surface. These forces include physical and chemical parts. The physical sorption is due to van der Waals attraction between adsorbate and adsorbent. The chemical sorption is due to the attractive chemical bonding between adsorbate and adsorbent (Zou 2020).

The capillarity is formed due to menisci between liquid water and gas mixtures. The existence of surface tension γ is the main cause of capillarity. The Young–Laplace gives the value of force balance at this interface:

$$P_c = P_g - P_w = \frac{2\gamma}{r^*} \tag{14}$$

where P_g and P_w are gas mixtures pressure and liquid–water pressure, respectively, and r^* indicates the mean radius of curvature.

The hysteresis between absorption and desorption isotherms is usually not very pronounced. Rode (1990) approximated the effect of hysteresis and found that the effect on the calculated water content results was not large. Currently still a large part of the heat and moisture models doesn't incorporate hysteresis and use the absorption isotherm or, where necessary, an average function of absorption and desorption.

Neglecting the hysteresis might not have a great influence on the water content but it dampens the fluctuations in relative humidity within the building assembly. In order to avoid this effect separate absorption and desorption isotherms and a validated method to interpolate between both curves must be employed (Trechsel 2001).

1.1.4 Mechanisms of Moisture Transport

Briefly, Trechsel (1994) described that moisture in the form of water vapor moves from one place to another either through mass transport, that is, by the movement of moist air, or through diffusion. The driving force of mass transport is air pressure; the driving force of diffusion is vapor pressure. The movement of liquid water can also result from wind pressure moving raindrops through cracks and joints, but as a general rule follows gravity forces.

Therefore, moisture transport in porous materials is very complex, which is a comprehensive expression of various mechanisms. As for weakly permeable porous media, it includes a series of phenomena, i.e. adsorption, desorption, condensation, evaporation, water flow, vapor diffusion, etc. Generally, the description of moisture transport can be divided into three stages according to the degree of relative humidity condition (Zou 2020).

According to the aforementioned transport mechanisms, it is possible to arrive at a definition of moisture transport as follows:

1. At high relative humidity stage, the water saturation is high which indicates most of the free liquid water is continuous. The transport of moisture is principally in the form of liquid water. The gradient of capillary pressure is the driving force for this movement, thus the extended Darcy's law can be applied.
2. At intermediate relative humidity stage, the region of continuous liquid–water continues to reduce which weakens the capillary transport. On the other hand, the vapor diffusion continues to strengthen. It means that the liquid water flow and vapor diffusion occur simultaneously. As the drying develops further, the vapor diffusion has preponderance compared with liquid water flow.
3. At low relative humidity stage, there is no continuous liquid water in the pore, while the gas mixtures are continuous. The vapor diffusion plays more important role in moisture transport. While, the contribution of liquid water flow is almost negligible. Indeed, the water molecules are adsorbed on the surface of pore walls

due to van der Waals forces (Iwamatsu and Horii 1996; Tuller and Or 2001). Evaporation occurs at the interface between liquid water and gas mixtures. This procedure is caused by the non-equilibrium between capillary pressure and vapor pressure. Then the evaporated vapor can be transported by diffusion through the pore. The diffusion includes ordinary diffusion, Knudsen diffusion (Mason 1983), and surface diffusion (Higashi et al. 1963).

- *Water transport in the liquid phase*

The water transport in the liquid phase can be processed through the phenomena of capillarity, thermal diffusion, electrokinetic, hydraulic flow and by the gravity effect. However, some of these phenomena can be neglected, such as thermal diffusion a very low rate of the total value of moisture transfer in buildings. Knowledge about the influence of electrokinetics is still very limited.

The influence of gravity on water transport is limited to large pore sizes ($>10^{-6}$ m) and since building materials mostly have smaller pores, this phenomenon can be neglected. However, in the liquid phase the phenomenon responsible for most of the moisture transport in porous materials is capillarity (Qiu 2003).

The capillarity phenomenon can be described by the capability of the liquid to pass through the porous medium without the help of external forces such as gravity, due to the attractive forces that are generated between the liquid and the solid material, overlapping the cohesive forces of the liquid and the gravitational action (Gonçalves 2007; Rouchier 2012).

Due to the complexity of the capillary network in a porous material to make it possible to perform an individual capillary analysis, the characterization of these materials regarding the transfer of water by capillarity, is performed at the macroscopic level through the global coefficients determination that are obtained in the capillarity test.

From Eqs. (15) and (16) it is possible to describe the amount of water absorbed and the water rise height at the scale of porous building materials. The total amount absorbed (W) and the height of capillary rise (H) are directly proportional to the square root of time. A ($\text{kg}/(\text{m}^2\sqrt{\text{s}})$) is the capillary water absorption coefficient and B ($\text{m}/\sqrt{\text{s}}$) is the capillary penetration coefficient. The initial values of w0 and h0 should also be observed, even if the specimens were oven dried.

$$W(t) = A \cdot \sqrt{t} + w_0 \tag{15}$$

$$H(t) = B \cdot \sqrt{t} + h_0 \tag{16}$$

The capillarity test is performed using a specimen made of a porous material, waterproofed on the lateral faces, in partial immersion, allowing capillary absorption to occur at its base (Fig. 3a) and from periodic weighings, the amount of water absorbed is determined, which in general is expressed through a graph similar to the one in Fig. 3b.

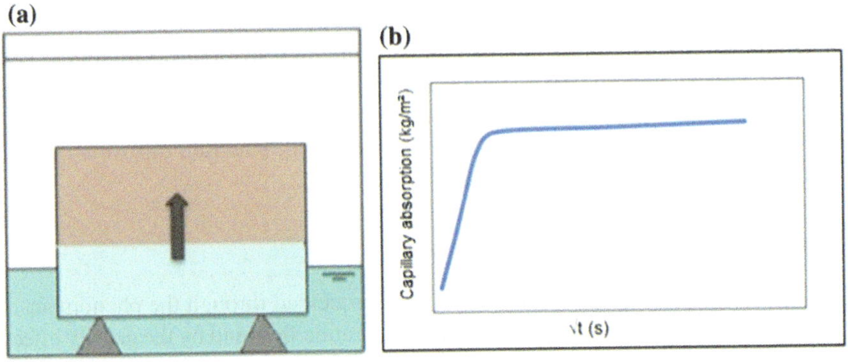

Fig. 3 **a** Capillarity absorption test and **b** Typical capillary absorption curve for a porous material

Therefore, water vapor diffusion implies the existence of a water vapor concentration gradient that promotes the flow of water molecules to places where the water vapor concentration is lower until a concentration homogenization is obtained.

- *Water transport in the vapor phase*

The water vapor transport in building materials is conditioned by diffusive processes and by convective movements inside the pores. The diffusive processes are essentially due to the existence of temperature gradients (thermal diffusion or Soret effect) and water vapor pressure gradients (gaseous diffusion proper). The existence of temperature gradients and convective movements are also usually neglected due to the low influence and the difficulty of determining the air pressure around the building and its reduced influence under normal conditions. However, diffusion becomes the phenomenon to promote most of the moisture transport in porous materials at that stage.

Therefore, water vapor diffusion implies the existence of a water vapor concentration gradient that promotes the flow of water molecules to places where the water vapor concentration is lower until a concentration homogenization is obtained. This phenomenon can be expressed by Fick's first law:

$$\overrightarrow{J_w} = -D_v \cdot \overrightarrow{\nabla c} \tag{17}$$

where, J_w is the vapor flux, in kg/m^2 s, D_v is diffusion coefficient, in m^2/s, and c represents the water vapor concentration, in kg/m^3.

The flow direction from the higher vapor concentration zone to the lower concentration zones is represented by the negative sign.

Assuming that air behaves as an ideal gas, the diffusion flux (c_w) can be expressed as a function of the water vapor pressure (p_w).

$$p_w = \frac{c_w \cdot R \cdot T}{M_w} \leftrightarrow c_w = \frac{p_w \cdot M_w}{R \cdot T} \tag{18}$$

where p_w is the water vapor pressure in Pa, c_w is the vapor concentration in kg/m3, R is the ideal gas constant, T is the temperature in Kelvin, M_w is the molar mass of water in kg/mol.

The water vapor permeability, π (kg/msPa), (Eq. 19), is a characteristic quantity of each material, and expresses the amount of vapor (kg) that passes through a material's unit thickness (m), per unit time (s) and surface area (m^2), when the pressure difference between the two material faces is also unit (Pa).

$$\pi = \frac{D_v \cdot M_w}{R \cdot T} \tag{19}$$

- *Moisture transport across interfaces*

The moisture transport across masonry is considered to be a governing parameter in several in several deterioration processes of masonry. Therefore, the study of moisture transport across the material interface is important to provide insight into the continuity between masonry wall layers.

Generically, three types of contact configurations can be considered in a multilayer wall, as illustrated in Fig. 4.

The three configuration types observed were defined by Freitas (1992) as follows:

- 'Hydraulic continuity (Perfect Hydraulic Contact)' when there is interpenetration of both layers' porous structure.
- 'Perfect contact' when there is contact without interpenetration of both layers' porous structure.
- 'Air space between layers' when there is an air pocket a few millimeters wide.

Hydraulic continuity (Perfect Hydraulic Contact) occurs in situations where the second layer is applied over the first layer and there is penetration of this material into the first layer, during the curing process of the second layer material. A practical example of this setting is the contact between the brick and the fresh mortar. Hydric resistance is defined as the greater or lesser ease of water diffusion.

However, Pel (1995) suggested that there is no perfect hydraulic contact between brick and mortar and that the moisture flux across the material interface is bound by maximum given by the hygric resistance of the interface (Freitas et al. 1991). Afterwards, several hypotheses were given for the origin of such a resistance, like an imperfect contact between material layers or the presence of an interface zone in the mortar layer.

The absorption behavior of brick/cement masonry composites with different interfacial configurations, was studied by Derluyn et al. (2011) and Janssen et al. (2012) in order to explain the imperfect hydraulic contact at the brick–mortar interface. Their studies confirmed that interfacial effects were proportional to the water extraction from the mortar during curing.

Other causes for hydraulic resistance at the brick–mortar interface could be the presence of air cracks resulting from damage (cracking) or due to poor workmanship during application of the fresh mortar (Groot and Gunneweg 2010). Brocken (1998)

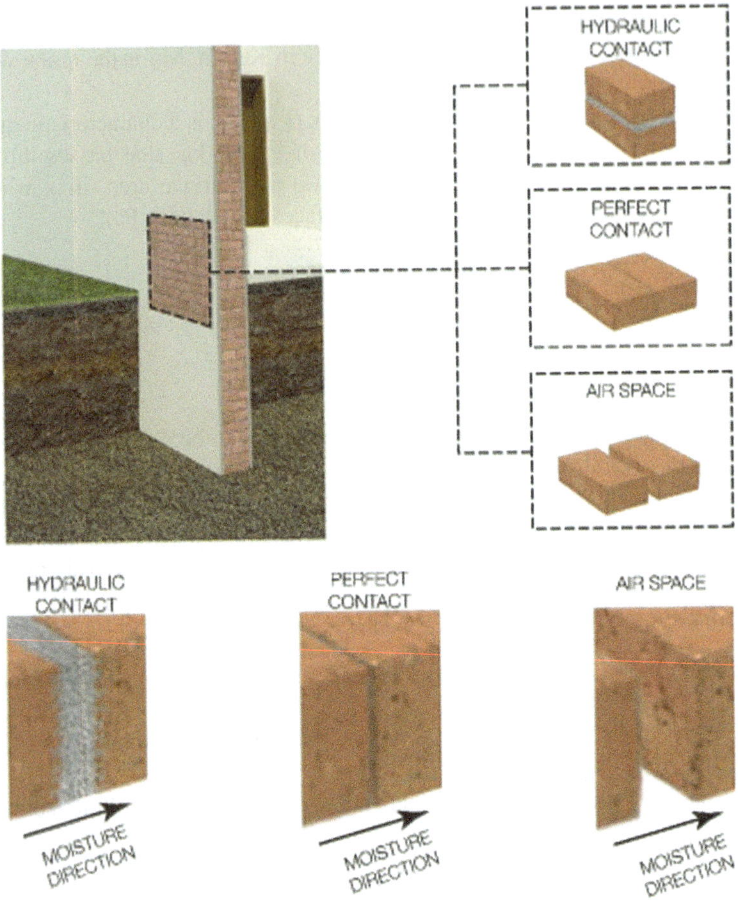

Fig. 4 Sketch of different contact configuration observed in building component

concludes in his study that even if the application and curing conditions were optimal, a certain hydraulic resistance is expected due to the discontinuity of the pore structure between the materials.

Perfect contact corresponds to cases where two layers are superposed; even if the contact surface is perfect, there is no continuity of the porous structure and, consequently, there is a hydric resistance that disrupts the transmitted flows.

Air space between layers reflects the case where the two layers have no physical contact and there is a few millimeters thick space separating them. It is an existing configuration at the interface between layers in the walls of some buildings.

However, few studies have been developed in this area, namely numerical simulation studies, which justifies the need for an extended study of moisture transport in multilayered walls, trying to understand the influence of the interface in this process.

1. Perfect Hydraulic Contact

A perfect hydraulic contact is characterised by the continuity of the macroscopic capillary pressure and the moisture flux across the interface (Freitas et al. 1996):

$$P_{c,i}(\theta_{l,i}) = P_{c,ii}(\theta_{l,ii}) \tag{20}$$

where the subscript i and ii refer to the material layers at both sides of the interface, P_c is the macroscopic capillary pressure of a pore, define as the pressure difference between the liquid phase and the gas phase ($P_c = P_l - P_g$) and θ_l is the volumetric liquid water content.

For different porous materials at both sides of the interface, these interface conditions result in a jump of the moisture content at the interface due to the difference in the moisture retention curve of the two materials.

$$\theta_{l,ii} = P_{c,ii}^{-1}(P_{c,i}(\theta_{l,i})) = f(\theta_{l,i}) \tag{21}$$

Assuming that hysteresis can be neglected and, consequently, that the capillary pressure curve is a single-valued function, this relation between the moisture contents of the two materials is unique.

The illustration of perfect hydraulic contact (hydraulic continuity) across the interface of two porous material layers with different capillary pressure curves is show in Fig. 5. The capillary pressure is continuous across the interface (Brocken 1998).

The moisture flux across such a material interface will be continuous:

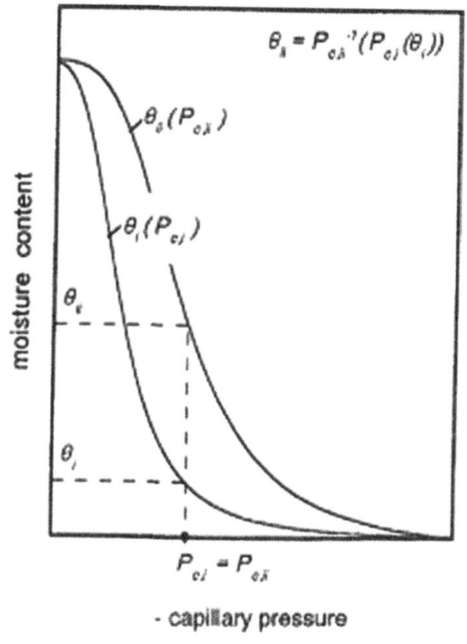

Fig. 5 Illustration of perfect hydraulic contact (hydraulic continuity) across the interface of two porous material layers with different capillary pressure curves

$$
\begin{aligned}
qi &= qii \\
\rho_l D_{\theta,i}(\theta_{l,i}) \frac{\partial \theta_{l,i}}{\partial x} &= \rho_l D_{\theta,ii}(\theta_{l,ii}) \frac{\partial \theta_{l,ii}}{\partial x}
\end{aligned} \tag{22}
$$

where q is the macroscopic isothermal liquid water flux (kg/m^2·s) by Darcy's law, as mentioned before, D_θ is the moisture diffusivity and ρ_l is the mass density of liquid water (kg/m^3).

If both materials are identical no influence of the interface will be observed when a water flow is crossing the interface and the governing moisture transport equation reduces to diffusion for a monolithic material.

The total mass balance for moisture transport combining the water vapor content θv with liquid water content θ_l, written as:

$$
\frac{\partial \theta}{\partial t} = \nabla (D_\theta \nabla \theta) \tag{23}
$$

where, $D_\theta = D_{\theta,v} + D_{\theta,l}$ and $\theta = \theta_v + \theta_l$.

If $\theta_l > \theta_v$, the contribution of water vapor transport to the total moisture transport can be negligible. However, for the dry state of a material, this condition is generally not satisfied.

Since hysteresis effects are neglected, the moisture diffusivity, D_θ, is a single valued and continuous function of the moisture content. Usually the differential can simply be solved with numerical algorithms applied for modeling unidirectional moisture transport for either water absorption or drying.

2. Imperfect Hydraulic Contact

An imperfect hydraulic contact between two porous materials in Brocken (1998) studies, have the interface (having no hygroscopic capacity) characterized by a so-called 'interface permeability'. This interface permeability is considered to be a macroscopic parameter which may originate either from a discontinuity of the pore structure of the materials at their interface, called by 'natural contact', from an air space between the material layers, or from a combination of both phenomena. Interface permeability was defined as:

$$
K_{if} = -\frac{q_{if}}{P_{c,ii}(\theta_{l,ii}) - P_{c,i}(\theta_{l,i})} \tag{24}
$$

where K_{if} is the interface permeability and, q_{if} the moisture flux across the interface. Water transport is assumed to be from material i to material ii. Hence, during the moisture transport, a jump in the capillary pressure will occur across the material interface:

$$
P_{c,ii}(\theta_{l,ii}) = P_{c,i}(\theta_{l,i}) - \frac{q_{if}}{K_{if}} \tag{25}
$$

As a result, the relation between the moisture contents of the two materials at their interface is not unique. Assuming that q_{if}/K_{if} is constant, two different relations exist depending on the direction of the moisture transport. From Fig. 6 it is possible to see that the relations between the moisture contents of material i and ii also depend on the trend of the capillary pressure curves of both materials. The resulting moisture across the material interface is written as:

$$q_i = q_{ii} = q_{if}$$
$$\rho_l D_{\theta,i}(\theta_{l,i}) \frac{\partial \theta_{l,i}}{\partial x} = \rho_l D_{\theta,ii}(\theta_{l,ii}) \frac{\partial \theta_{l,ii}}{\partial x} = K_{if}\left(P_{c,ii}(\theta_{l,ii}) - P_{c,i}(\theta_{l,i})\right) \tag{26}$$

3. Air Space Between Layers

An air space between two porous layers has the isothermal moisture flux over this space, with stagnant air, written as (Brocken 1998):

$$q_{if} = -\delta_a \frac{\Delta p_v}{d} = -\delta_a p_{vs} \frac{\Delta h}{d} \tag{27}$$

where, δ_a is the water vapor permeability, d is the thickness of the air space, h is the relative humidity, p_v and p_{vs} represent the vapor pressure and the saturation vapor pressure respectively, and v_l and v_v are the molar volume of liquid water and water

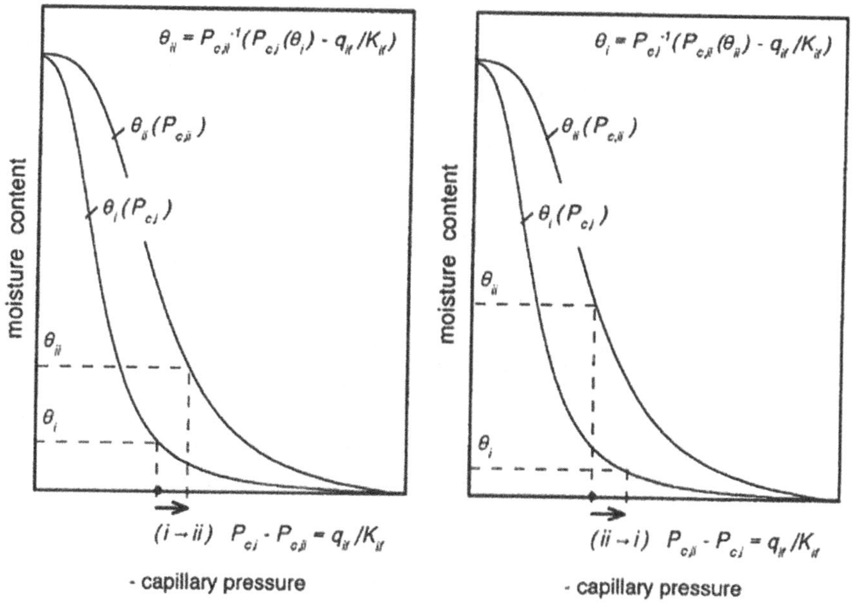

Fig. 6 Illustration of imperfect hydraulic contact (hydraulic discontinuity) across the interface of two porous material layers with different capillary pressure curves

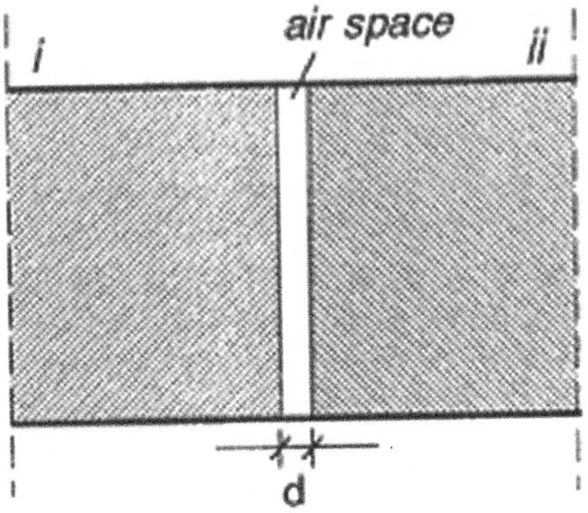

Fig. 7 Schematic representation of imperfect hydraulic contact resulting from an air space of thickness d between materials layers

vapor respectively. The flux moisture flux can be rewritten using Kelvin's Eq. (11):

$$q_{if} = -\frac{\delta_a p_{vs}}{d}\left(\exp\left(\frac{v_l P_{c,ii}(\theta_{l,ii})}{RT}\right) - \exp\left(\frac{v_l P_{c,i}(\theta_{l,i})}{RT}\right)\right) \tag{28}$$

The interface permeability resulting from the air space configuration (Fig. 7) will depends on the thickness of the air space between the material layers and the trend of the capillary pressure curves is variable. Equation (29) show the interface permeability resulting from an air space between two material layers, where the thickness of the air space is a constant value.

$$K_{if} = -\frac{\frac{\delta_a p_{vs}}{d}\left(\exp\left(\frac{v_l P_{c,ii}(\theta_{l,ii})}{RT}\right) - \exp\left(\frac{v_l P_{c,i}(\theta_{l,i})}{RT}\right)\right)}{P_{c,ii}(\theta_{l,ii}) - P_{c,i}(\theta_{l,i})} \tag{29}$$

An air space between two porous layers leads to a discontinuity in the moisture profile by a one-dimensional free water imbibition test (Bear and Bachmat 2012): The free water level in contact with the first layer and the moisture content will evolve to the capillary moisture content of the porous material while the moisture content of the other layer will evolve to a hygroscopic moisture equilibrium by water vapor transport over the air layer interface from the first to the second material layer.

2 Moisture Transfer Computer Simulation Programs

Moisture problem is one of the most important factors related to the performance of the building (Guimarães et al. 2018). Evaluating the moisture behavior in the building envelope is important to avoid the damage caused by moisture accumulation in the building envelope which can cause structural and physical problems for the building.

The first models developed to analyze moisture transport were focused on the analysis of porous soils. Lewis (1921), Richards (1931), Phillip and De Vries (1957) and Luikov (1966) elaborated the first phenomenological models to characterize transport in unsaturated porous media. The first coupled HAM model was instituted by Philip and Vries (1957). The model has as a function of temperature soil moisture transport and the main driving forces considered for its penetration are temperature and volumetric water content. Luikov's model presented a descriptive study for the estimation of MAP transport on the macroscopic scale in capillary porous media. Whitaker (1977) developed a detailed theory describing each phase of the transport equations in the year 1977. This assisted in the development of the first technique designed to assess moisture in building materials in the 1980s, known as the Glaser method. Pedersen (1992) and Kunzel (1995) developed more complete models that take into account the transport of liquids and diffusing vapors. The model is based on Kiebl's theorem, and was based on experimental work to simplify the models and the determination of the transfer coefficients. Mendes et al. (1999) developed a model based on the Philip and DeVries model. The developed model predicts heat and moisture transfer through porous building elements as well as complex configuration cases such as multilayer walls (Mendes et al. 2002).

Hygrothermal simulations in the field of building physics are widely used to predict the hygrothermal performance of building materials, components, and entire buildings. The available tools vary in their degrees of mathematical sophistication and runtime requirements, i.e., based on different mathematical models (physical descriptions), use different driving potentials, and use different numerical methods for space and time discretization. Thus, the tools have different potentials, strengths and weaknesses, for example, the ability to include air transfer (Langmans et al. 2012), 2D or 3D phenomena (Ruisinger and Kautsch 2020), or the ability to simulate a large number of zones in a reasonable runtime (Karoglou et al. 2007).

The main difference considered in studying moisture transport with hygrothermal models for porous building materials and envelopes is the moisture migration conduction potential. The moisture conduction potentials used in hygrothermal models in the existing literature include water vapor partial pressure, moisture content, capillary pressure, relative humidity, and air moisture content (Chang et al. 2020). This variety of models encompassing different methods and input and output parameters makes it challenging to seal the most suitable tool for studying a specific problem.

Computational fluid dynamics (CFD) is the main numerical simulation approaches used in the moisture transport analyses. The CFD approach, focuses on the numerical meshing of the entire building component volume into small elements or volumes, which results in a large computing load, particularly for unsteady states simulations.

The mathematical model of this system is consist in a set of partial differential equations (PDEs) for describing the laws of physics (the conservation of momentum, mass, and energy) for the space and time-dependent descriptions (Hens and Heat 2002).

The mathematical model can consist of one or several PDEs (describing the relevant laws), together with boundary and initial conditions. Normally, the right-hand side of the PDE represents the transfer of heat and moisture, as quantified by different material properties and different potentials. The left-hand side represents the storage (Hens 2007). The solution to the PDEs is represented by dependent variables (e.g. temperature fields, RH fields, or velocity fields) described in space and time along the independent variables x, y, z, and t.

For building envelopes, the current challenge for hygrothermal models is to consider the discontinuity at the interface between material layers in multilayered walls.

2.1 Simulation Programs with Wider Applicability

In a recent study by Verma et al. (2022) was highlighted the utilization of some major simulation tools in the literature studied for hygrothermal dynamics for developing energy-efficient buildings, considering building materials (Fig. 8).

An accurate hygrothermal model is essential to improve and evaluate the moisture transport. The types model interesting to know in this is study are based on physical knowledge of the system and energy balance equations. These are often obtained through energy simulation software such as TRNSYS, EnergyPlus, WUFI, COMSOL Multiphysics. However, as all model, for such models exists disadvantages. Benzaama et al. (2020) worked out a comparative study and presented drawbacks like: (i) such models often require a great large set-up and computation time. (ii) involve a lot of inputs to define the model, such as the composition of the building envelope. In some studies, it is difficult, if not impossible, to recover this input.

The commercially available tools WUFI® Pro and 2D [applying the PDEs described in Künzel and Kiesel (1997)] is widely used by researchers to investigate the heat and moisture performance of building components.

- **WUFI** was the program used in this study and will be described extensively in Sect. 2.3 as it being the specialized tool for studying hygrothermal performance is used most, while Ansys and EnergyPlus due to their wider range of applicability

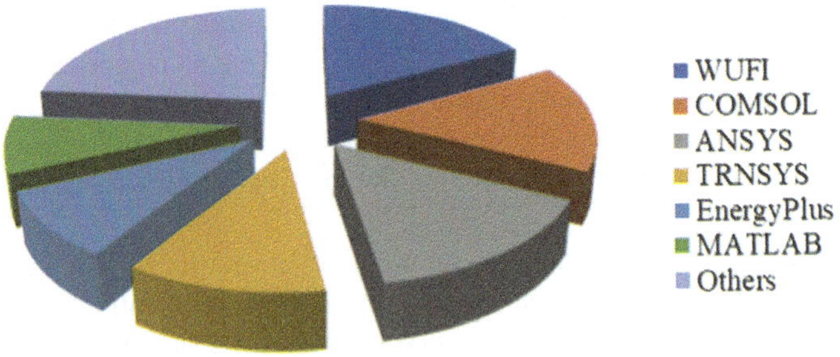

Fig. 8 Utilization of different tools in hygrothermal analyzing

are gaining the attention of users. However, there is a diverse range of tools and each of them is having typical utilization.

- **COMSOL** Multiphysics® is a simulation platform that provides fully coupled multiphysics and single-physics modeling capabilities. Its originates from partial differential equation (PDE) Toolbox of MATLAB. Since it officially named COMSOL Multiphysics in 2003, it has been absorbing new calculation methods and techniques, and also extending new application modules. The software based on finite element analysis, which has a large set of functions for analyses and solution. It includes heat transfer module, electromagnetic module, acoustics module, earth science module, chemical engineering module and structural mechanics module. The platform provides two kinds of operation modes, graphical user interface style, and command style by creating scripts. Both modes provide convenience for users mostly (Liang and Chen 1999). Script mode is mainly for optimum design and second development for COMSOL Multiphysics. The software includes three sections. Pre-process, solution, and post- process. Creating finite element model and setting load parameters are belong to pre-processing. Mesh division and solving equations are all belong to solution section. Results visualization and analysis are belong to post- processing (Wang et al. 2011). The Model Builder includes all of the steps in the modeling workflow—from defining geometries, material properties, and the physics that describe specific phenomena to solving and postprocessing models for producing accurate results. According to the comsol.com/comsol-multiphysics website.
- **ANSYS** is a general-purpose finite-element modeling package for numerically solving a wide variety of mechanical problems. These problems include static/dynamic, structural analysis (both linear and nonlinear), heat transfer, and fluid problems, as well as acoustic and electromagnetic problems. In general, a finite-element solution may be broken into the following three stages (Nakasone et al. 2006): (i) Preprocessing: defining the problem—The major steps in preprocessing are: (i) define keypoints/lines/areas/volumes, (ii) define element type and material/geometric properties, and (iii) mesh lines/areas/volumes as required. The amount of detail required will depend on the dimensionality of the analysis, i.e., 1D, 2D, axisymmetric, and 3D. (ii) Solution: assigning loads, constraints, and solving. Here, it is necessary to specify the loads (point or pressure), constraints (translational and rotational), and finally solve the resulting set of equations. (iii) Postprocessing: further processing and viewing of the results. In this stage one may wish to see (i) lists of nodal displacements, (ii) element forces and moments, (iii) deflection plots, and (iv) stress contour diagrams or temperature maps.
- **TRNSYS** (Klein et al. 2007) is a transient system simulation environment with a modular structure that allows independent components, consisting of validated mathematical models representing individual components in an energy system, to be joined together based on (real life) scenarios to produce an output based on the model requirements. TRNSYS also allows the user to link to external programs for further analysis of results (e.g. Microsoft Excel, MATLAB). According to the trnsys.com website. TRNSYS is made up of two parts. The first is an engine (called the kernel) that reads and processes the input file, iteratively solves the

system, determines convergence, and plots system variables. The kernel also provides utilities that (among other things) determine thermophysical properties, invert matrices, perform linear regressions, and interpolate external data files. The second part of TRNSYS is an extensive library of components, each of which models the performance of one part of the system. The standard library includes approximately 150 models ranging from pumps to multizone buildings, wind turbines to electrolyzers, weather data processors to economics routines, and basic HVAC equipment to cutting edge emerging technologies. Models are constructed in such a way that users can modify existing components or write their own, extending the capabilities of the environment. Steeman et al. (2010) used the coupling heat, air and moisture transfer (HAM) in porous materials within the Building Energy Simulation tool TRNSYS with implicit time discretization scheme. The developed coupled model is flexible and is able to simulate multilayered walls with variable boundary conditions in real building application modeling. Building energy simulation softwares like TRNSYS and EnergyPlus are mainly used to simulate temperature variations and energy demands in specific spaces at a large scale (Djedjig et al. 2015). As a result, moisture exchange models at the wall scale were used in these tools using a simplified model that neglects the coupling of heat and moisture transfer phenomena through the building envelope (Preuss 2015).

- **EnergyPlus** is a whole building energy simulation program that engineers, architects, and researchers use to model both energy consumption—for heating, cooling, ventilation, lighting and plug and process loads—and water use in buildings. Some of the notable features and main capabilities of EnergyPlus include: Heat balance-based solution of radiant and convective effects that produce surface temperatures, thermal comfort, and condensation calculations; Combined heat and mass transfer model that accounts for air movement between zones; Functional Mockup Interface import and export for co-simulation with other engines; Transient heat conduction through building elements such as walls, roofs, floors, etc. using conduction transfer functions; Thermal comfort models based on activity, inside dry-bulb temperature, humidity, etc. (Crawley et al. 2001).
- **MATLAB** is a high-level technical computing language and interactive environment for algorithm development, data visualization, data analysis, and numerical computation. Using MATLAB, you can solve technical computing problems faster than with traditional programming languages, such as C, C++, and Fortran. Millions of engineers and scientists worldwide use MATLAB® to analyze and design the systems and products. The matrix-based MATLAB language is the world's most natural way to express computational mathematics. Built-in graphics make it easy to visualize and gain insights from data. The desktop environment invites experimentation, exploration, and discovery. These MATLAB tools and capabilities are all rigorously tested and designed to work together. MATLAB code can be integrated with other languages, enabling the users to deploy algorithms and applications within web, enterprise, and production systems. According to the mathworks.com website. Among the other programs used, the following have

been chosen as interest programs that are commonly used in moisture transfer studies.

- **Delphin 5** is a commercial two-dimensional numerical simulation program for the transport of heat, air, moisture, pollution and salt transport in porous building materials. This program can be used to simulate mass and energy transport processes for user-defined climatic conditions or real climates (temperature, relative humidity, incident rainfall, wind speed and direction, longwave and shortwave radiation) (Nicolai et al. 2007).
- **MOISTURE-EXPERT** is a one or two-dimensional numerical modeling software of heat, air and moisture transport in building construction systems. The program is basically software developed by Oak Ridge National Laboratory and the Fraunhofer Institute for Building Physics to adapt the original European version of the WUFI software for the US and Canada. The model handles vapor and liquid transport separately. The moisture transport potentials are vapor pressure and relative humidity, and the energy transport potential is temperature. The model includes the ability to handle absorption and adsorption isotherms, temperature dependent, and transport properties as a function of drying or humidification processes. It can be considered as a disadvantage for use, but as an advantage in the accuracy of the results, the fact that the program is highly complex, typically requiring more than 1000 inputs for the one-dimensional simulations. The inputs include outdoor weather conditions, indoor weather conditions, material properties, and system characteristics (Karagiozis 2001).

2.2 Common Limitations in Mathematical Models

Engineers who use finite element analysis must understand the limitations of the finite element procedures. There are numerous circumstances that can degrade the quality of a calculation or even render it worthless.

There are various sources of error that can contribute to incorrect results ["Finite element analysis—theory and application with ANSYS" (1999)]. They include:

1. Wrong input data, such as physical properties and dimensions. This mistake can be corrected by simply listing and verifying physical properties and coordinates of nodes or keypoints (points defining the vertices of an object) before proceeding any further with the analysis.
2. Selecting inappropriate types of elements. Understanding the underlying theory will benefit you the most in this respect. You need to fully grasp the limitations of a given type of element and understand to which type of problems it applies.
3. Poor element shape and size after meshing. This area is a very important part of any finite element analysis. Inappropriate element shape and size will influence the accuracy of your results. It is important that the user understands the difference between free meshing (using mixed-area element shapes) and mapped meshing (using all quadrilateral area elements or all hexahedral volume elements) and the limitations associated with them.

4. Applying wrong boundary conditions and loads. This step is usually the most difficult aspect of modeling. It involves taking an actual problem and estimating the loading and the appropriate boundary conditions for a finite element model. This step requires good judgment and some experience.

Many hygrothemal analysis tools require a building model to be created specifically for that tool, often graphically but sometimes via text input (Murray et al. 2009). The complex hygrothermal processes in a building component need to be simplified to make their simulation accessible to present-day computers.

With the vast number of hygrothemal analysis tools available and with their varying capabilities, it is often to find studies using different tool because of the limitations found according to the models. Any software has its share of limitations, requires a certain skill level, and the user must be aware of what the model can and cannot do. Common limitations in mathematical models include:

- Effects associated with phase change, liquid to ice, are neglected.
- Climatic load due to driving rain is simplified; Wind-driven rain is an approximation of surface wetting.
- No hysteresis is accounted for; Hysteresis of the moisture retention curve is not taken into account.
- No chemical reactions are considered.
- Ageing effects or changes in the geometrical dimensions are neglected.
- Deformation of the porous structure caused by the ice content changes is neglected.
- The interface between two capillary-active materials is treated as ideally.
- No user-friendly interface for inputting data.
- Limited material properties are currently available; A limited number of laboratory benchmarking tests have been performed, none with field data.

2.3 WUFI-2D

For the purpose of this work, the moisture performance simulation model WUFI 2D 4.3, which is a windows based program for the hygrothermal analysis of building envelope and components constructions, is selected. WUFI (Warme und Feuchte instationar—Transient Heat and Moisture) is a one-dimensional, two-dimensional or 3D model for heat and moisture transport developed by the Fraunhofer Institute in Building Physics (IBP) in Holzkirchen, Germany. Which can be used to assess the heat and moisture distributions for a wide range of building material classes and climatic conditions.

The program requires only the standard material properties and allows the determination of moisture storage and liquid transport functions. It was validated through in situ and laboratory data, allowing a realistic simulation of the hygrothermal behavior of building elements, monolithic or multilayer, exposed to real climatic conditions.

The two-dimensional analysis performed by the WUFI-2D program is based on the finite-volume method, allowing the modeling of building elements with complex geometries (Freitas et al. 2008).

The mathematical model used in WUFI was developed by Kunzel (1995) based on Kiebl's theorem. In this model the non-steady heat and moisture transport processes in building components are described by the following coupled differential equations for heat transport and moisture transport:

$$\frac{\partial H}{\partial v}\frac{\partial v}{\partial t} = \frac{\partial}{\partial x}\left(\lambda\frac{\partial v}{\partial x}\right) + h_v\frac{\partial}{\partial x}\left(\frac{\delta}{\mu}\frac{\partial p}{\partial x}\right) \tag{30}$$

$$\rho_w\frac{\partial u}{\partial \varphi}\frac{\partial \varphi}{\partial t} = \frac{\partial}{\partial x}\left(\rho_w D_w\frac{\partial u}{\partial \varphi}\frac{\partial \varphi}{\partial x}\right) + \frac{\partial}{\partial x}\left(\frac{\delta}{\mu}\frac{\partial p}{\partial x}\right) \tag{31}$$

where, Dw [m^2/s] is the liquid transport coefficient, H [J/m^3] is the enthalpy of moist building material, hv [J/kg] is the evaporation enthalpy of water, p [Pa] is the water vapor partial pressure, u [m^3/m^3] is the water content, δ [kg/msPa] is the water vapor diffusion coefficient in air, v [°C] is the temperature, λ [W/mK] is the heat conductivity of moist material, μ [–] is the vapor diffusion resistance factor of dry material, ρw [kg/m^3] is the density of water and φ [–] is the relative humidity

The left-hand sides of both equations consist of the storage terms. Heat storage comprises the heat capacity of the dry material and the heat capacity of the moisture present in the material. Moisture storage is described by the derivative of the moisture storage function mentioned above. On the righ-hand side of the equations we find the transport terms. Heat transport is the sum of moisture-dependent thermal conductivity and vapor enthalpy flow. This heat transport by vapor enthalpy flow is due to water evaporating in one place and thereby absorbing latent heat from this place, and then diffusing to a different place, condensing there and releasing latent heat. This kind of heat transport is often called latent heat effect. Liquid transport (through surface diffusion and capillary conduction, both due to a gradient of relative humidity) shows only a relatively minor temperature dependence. Vapor diffusion, on the other hand, is strongly affected by the temperature field, since the saturation vapor pressure increases exponentially with temperature. The differential equations are discretised by means of an implicit finite volume method and are iteratively solved according to the scheme described by the flow chart shown in Fig. 9. The accuracy of the numerical solution depends on the mesh widths of the numerical grid, the size of the time steps and the choice of the convergence criteria. Usually the numerical solution is sufficiently accurate, so that the effect of numerical parameters can be ignored in comparison with the effects of the physical parameters like material and climate data. After the calculation the result should be critically assessed in order to exclude user errors or severe convergence errors. Convergence errors are indicated by WUFI, and their effect is assessed by a comparison of the sum of the moisture flows with the water accumulated in the component. False input or unrealistic material data can only be controlled by plausibility checks. According to the wufi-wiki website.

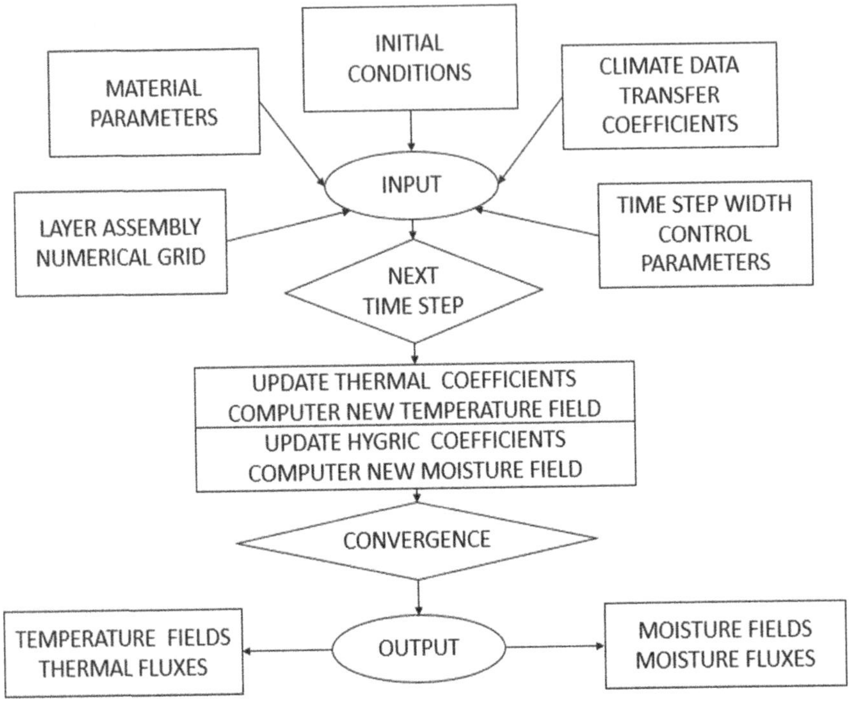

Fig. 9 Flow chart for the WUFI model

3 Numerical Simulation

In the previous chapter, it was possible to understand the functionality of the WUFI-2D program that will be used in this study to simulate moisture transport in samples with different contact configuration between the materials. Characteristic information of the simulations performed in this study was also provided.

The simulations performed in this work aim to analyze the influence of the interface in the different types of contact—perfect hydraulic contact between mortar and brick and the air gap between brick layers—on moisture transport, through the verification of moisture content profiles obtained with the simulations performed. Also, it is intended to briefly compare the results of moisture content profiles obtained in the experimental study of Azevedo (2019) and the numerical simulations performed in WUFI-2D, in order to contribute to the experimental validation. For this purpose, the same material properties, initial conditions and boundary conditions of the experimental study were used in the execution of the simulations.

As mentioned earlier, the mathematical model used in WUFI-2D has as a potential conductor for the capillary moisture transport the relative humidity, which is continuous in multi-layered building components, i.e. continuous across material independent material boundaries. This is why the hydraulic contact configuration is considered as perfect, with no consideration of the second layer material penetrating into the first layer and the porous network discontinuity at the interface, thus treated as ideal between two active capillary materials for the contact condition of the configurations in the simulations performed.

Although the mechanisms of moisture transport in a single building material have been and continue to be widely studied, the hydraulic characteristics of the interface at different types of contacts between materials are still poorly understood and, for this reason, the simplified assumption of perfect contact is widely used in hygrothermal models. However, compared to a monolithic element, the multilayer element exhibits delayed liquid transport across the interface between the layer materials (Azevedo 2019).

The geometry and properties of the samples to be analyzed took into consideration two types of red ceramic bricks: Brick A and Brick B; two types of mortars: Cementitious and Lime; three configurations for the application of hydraulic joint for each type of brick and each type of mortar; and six configurations for air gap application between brick layers for each type of brick. The simulations of monolithic bricks samples are used as a basis for comparing the effects caused by the interface on the samples.

The configurations of the simulated models are presented in Figs. 10 and 11 and Table 2, the properties and initial conditions of the materials used in the simulation are presented in Tables 3 and 4, respectively.

For the water absorption simulation on the samples with perfect hydraulic contact of mortar (cement and lime) with 2 cm, 5 cm and 7 cm distance from the base of the sample (Fig. 10), a 1 cm thick layer of each material was inserted between the two blocks. The adiabatic system was applied to the lateral surfaces, as the goal is to analyze moisture transport on a single surface and avoid heat and moisture exchange at the borders. The water available for absorption is taken from an exterior climate file with rain of 1000 Ltr/m^{2}h, at a 20 °C temperature and 100% relative humidity. The program calculates the water content absorbed from the surface by measuring the rainfall incident on the surface every one hour. For the top surface an indoor climate with 20ºC and RH 50% was applied according to the laboratory conditions for the experimental test (see Fig. 11).

The properties of air layers must be described by sets of material parameters which had been intended to describe porous materials. The effective material parameters of the air layers describe total heat transport and total vapor transport (due to diffusion and free convection) in an unventilated air layer between non-metallic surfaces. Since the percentage of convection in the total transport is affected by the thickness of the air layer, the effective material parameters also depend on the layer thickness. Air layer thicknesses not provided in the database can be created with the method discussed in the online help.

Currently the program has in the database new air layers "without additional moisture capacity". Their free saturation of 17 g/m^3 corresponds to the saturation moisture of air at 20 °C. However, the free saturation is always fixed at the value cited and does not vary with temperature, as it would in real air. The moisture contents

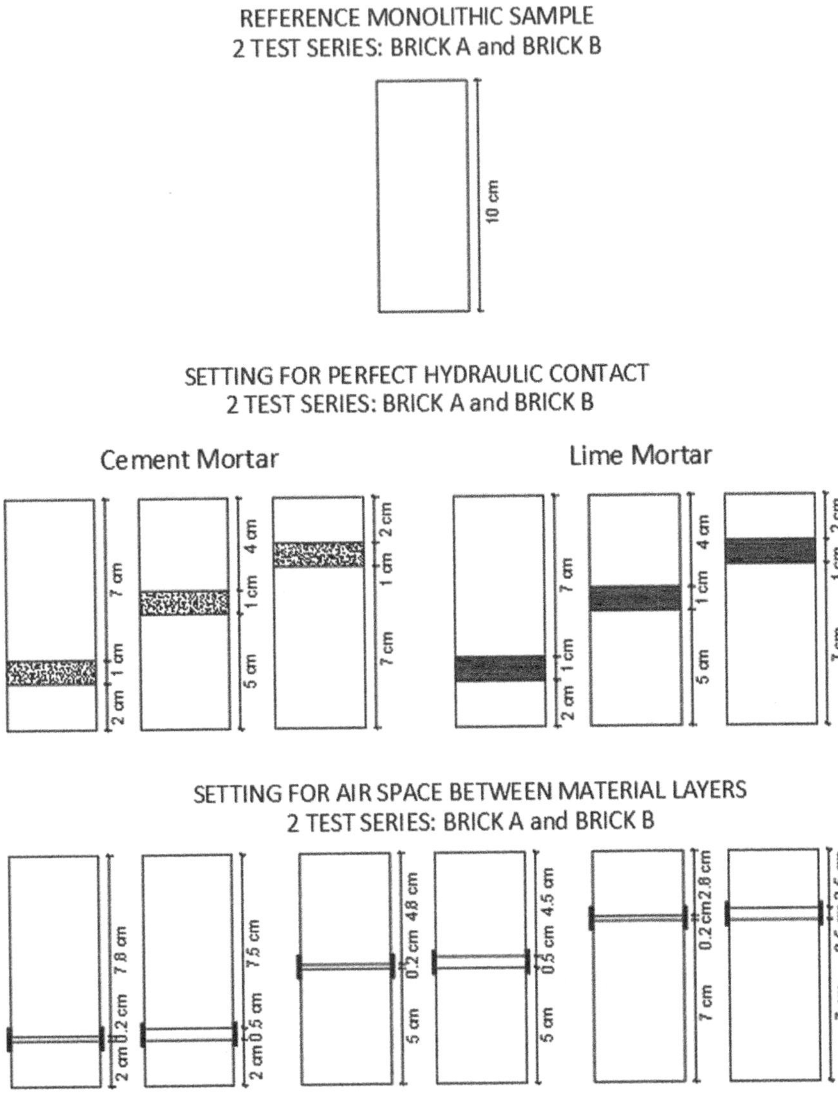

Fig. 10 Simulation setup: **i** monolithic block; **ii** a block with cement and lime mortar interface at 2 cm, 5 cm and 7 cm; **iii** an air space with 0.5 cm and 0.2 cm cavity at 2 cm, 5 cm and 7 cm

and hygroscopic inertia of these air layers are at a realistic level, but can prove numerically challenging.

The simulations were performed for the same time period, 72 h. These parameters were used in the Azevedo (2019) experimental trial.

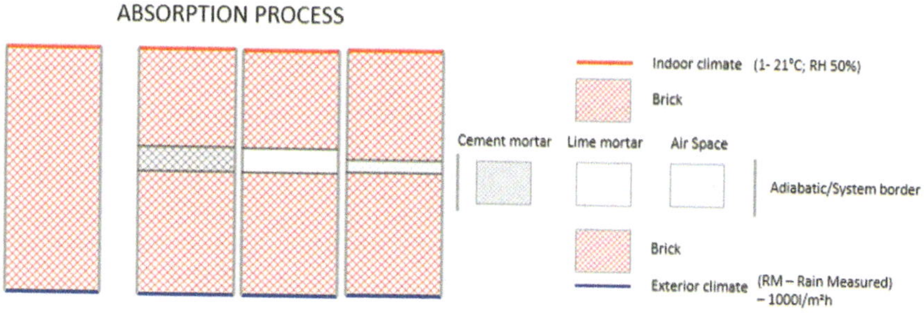

Fig. 11 Settings for the simulations processes

4 Results and Discussions

4.1 Moisture Transport Across Hydraulic Contact Interface—Wetting Process

The simulation of moisture transport along ceramic brick specimens with perfect hydraulic contact and air space interface between material layers were performed. The objective is to analyze the moisture behavior in monolithic samples, multilayer samples of different materials and samples with air space between ceramic material layers. By comparing the monolithic samples with the multilayer samples, it is possible to evaluate the influence that different types of contact between different materials have on moisture transport in multilayer components.

As stated in the methodology, this work also performs comparisons with the experimental survey results of Azevedo (2019), in order to validate the experimental result and contribute to the evaluation of the tool in real applications.

However, since it is considered that the existence of the interface between the materials layers generates a hydric resistance in the water transport, the simulations performed analyze the transport of the moisture front in immediately after the first change point in comparison with the monolithic sample.

Although the program considers a continuous transport across the interface between the materials, it is still possible to analyze a discontinuity caused by the difference in materials. This result will be important to validate the experimental results that identifies the discontinuity at the interface and which contributes to higher water resistance values. The method used to measure the water resistance values for the simulations was the same method used for Azevedo (2019) to obtain the values through the experimental gravimetric method that uses the following equation:

$$RH = \frac{\Delta M_w}{\Delta t} \tag{32}$$

Table 2 Settings for the simulations

Material	Wetting environment 1 (21 °C, RH 50%)		
	Monolithic	Perfect hydraulic contact	Air space
Red brick type A 40 × 100 mm^2		Cement mortar at h = 20 mm	Cavity of 2 mm at h = 20 mm
		Cement mortar at h = 50 mm	Cavity of 5 mm at h = 20 mm
		Cement mortar at h = 70 mm	Cavity of 2 mm at h = 50 mm
		Lime mortar at h = 20 mm	Cavity of 5 mm at h = 50 mm
		Lime mortar at h = 50 mm	Cavity of 2 mm at h = 70 mm
		Lime mortar at h = 70 mm	Cavity of 5 mm at h = 70 mm
	1 sample	6 samples	6 samples
Red brick type B 50 × 100 mm^2		Cement mortar at h = 20 mm	Cavity of 2 mm at h = 20 mm
		Cement mortar at h = 50 mm	Cavity of 5 mm at h = 20 mm
		Cement mortar at h = 70 mm	Cavity of 2 mm at h = 50 mm
		Lime mortar at h = 20 mm	Cavity of 5 mm at h = 50 mm
		Lime mortar at h = 50 mm	Cavity of 2 mm at h = 70 mm
		Lime mortar at h = 70 mm	Cavity of 5 mm at h = 70 mm
Wetting dry	1 sample	6 samples	6 samples
	2 samples	12 samples	12 samples
	2 samples	24 samples	12 samples
	4 simulations	36 simulations	24 simulations
Total	64 simulations		

where Δt (s) and ΔM_w (kg/m^2) are the variation of the time and water absorption immediately after the knee point, respectively. Nevertheless, the program generates graphs on moisture content (kg/m^3) by time in hours (h).

The program interprets the hydric resistance that limits the moisture transport through the different properties on contact between different materials. The hydric resistance (RH) is the measurement calculated experimentally (during a water absorption test) by the slope of the curve of mass change as a function of time, after the

Table 3 Hygrothermal properties of materials

Properties	Materials/layers					
	Red brick type A	Red brick type B	Cement mortar	Lime mortar	Air space 2 mm	Air space 5 mm
Bulk density, ρ [kg/m^3]	1800	1600	1878	1810	12.9	12.9
Porosity, ε [m^3/m^3]	0.32	0.38	0.20	0.21	0.999	0.999
Specific heat capacity, C [J/kgK]	850	850	900	900	1000	1000
Thermal conductivity (Dry), λ [W/mK]	0.40	0.38	0.786	0.799	0.0125	0.02778
Water vap. diff. resist. factor, μ [–]	33.1 (dry cup)	21.4 (dry cup)	23.90	14.25	5.59	3.224
Moisture content, w80% [kg/m^3]	w80% = 15.51	w80% = 9.75	w80% = 38.16	w80% = 8.16	0.0136	0.0136
Moisture content, wsat [kg/m^3] Free water saturation	wsat = 261.38	wsat = 233.07	wsat. = 228.59	Wsat. = 186.61	0.017	0.017
Water absorption coeff. [kg/m$^2\sqrt{s}$]	0.10	0.19	0.15	0.12	–	–

knee point (Fig. 12). The measured values of the hydric resistance for the simulations followed the same methodology, using the variation curve of water content as a function of time after the slope point that intersects the moisture transport curve for the monolithic sample.

Figures 13 and 14 show the plot of the simulations after 72 h for the moisture transport (kg/m^3) per hour along the samples with perfect hydraulic contact between brick and mortar layer, comparing with the moisture transport behavior in the monolithic brick sample. Also shown is the zoom in at the time when the multilayered sample curves intersect the monolithic sample curve and reduce the transported moisture content.

As reported previously, for the perfect hydraulic contact analysis two configurations are shown: The first configuration has the cement mortar between two layers of brick and has three samples with different interface location height (2 cm, 5 cm, 7 cm) and the second configuration has the lime mortar between two layers of brick and also has three samples with different interface location height (2 cm, 5 cm, 7 cm).

Table 4 Initial conditions of materials/layers

Initial conditions			
	Temperature (°C)	Moisture content, w (kg/m^3)	Relative humidity, RH (%)
Red brick	20 °C	w80% = 15.51	0.8
type A		wsat = 261.38	1
Red brick	20 °C	w80% = 9.75	0.8
type B		wsat = 233.07	1
Cement mortar	20 °C	w80% = 38.16	0.8
		wsat. = 228.59	1
Lime mortar	20 °C	w80% = 8.16	0.8
		wsat. = 186.61	1
Air space	20 °C	w80% = 0.0136	0.8
		wsat. = 0.017	–

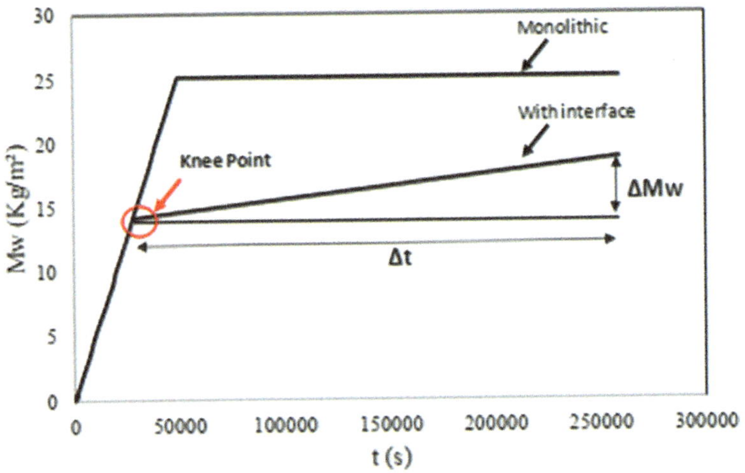

Fig. 12 Hydric resistance (RH) in gravimetric method used to the simulations results

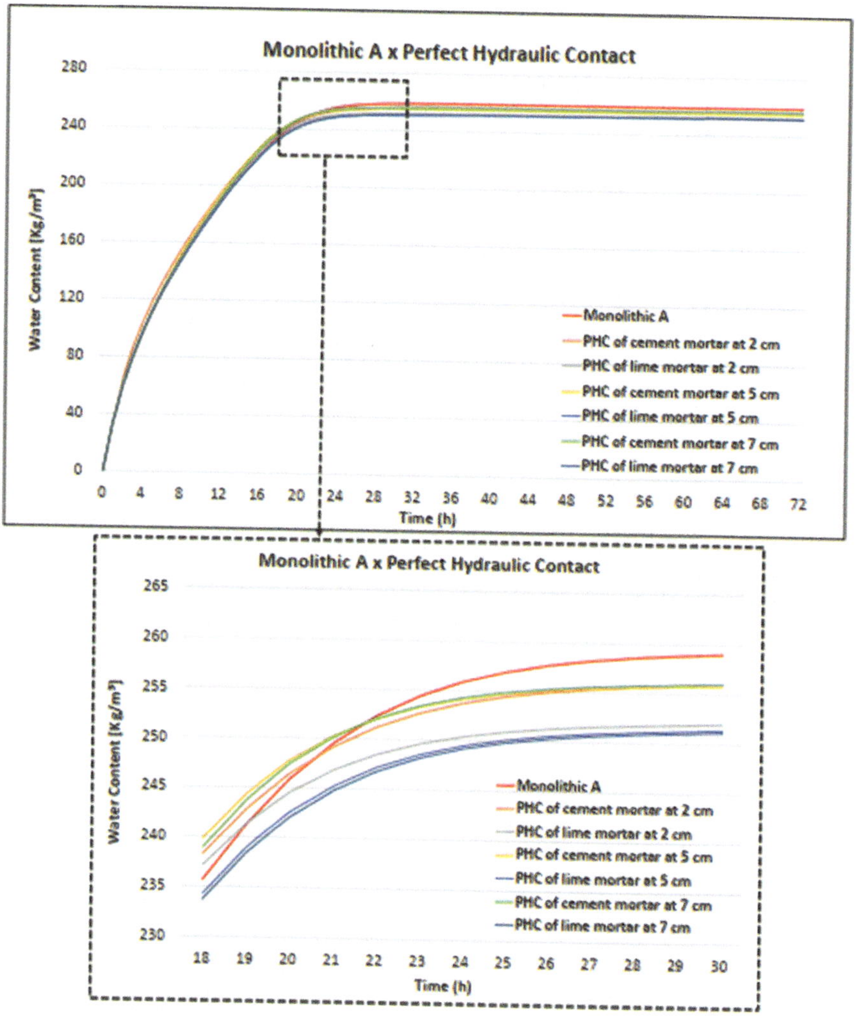

Fig. 13 Water content graph along brick A after 72 h of simulated transport for perfect hydraulic contact (cement × lime)

Through the results obtained in the graphs (see Figs. 13 and 14), it was observed that there is a reduction in the water content transported for the samples with perfect hydraulic contact in relation to the transport on the monolithic sample. For samples with lime mortar there was a greater reduction in the amount of water content absorbed compared to the samples with cement mortar. The explanation for this result is directly related to the materials' water storage capacity (saturation point).

Figures 13 and 14 indicate a decrease in the water absorption rate after the intersection point in the water content curve for the monolithic sample. As explained

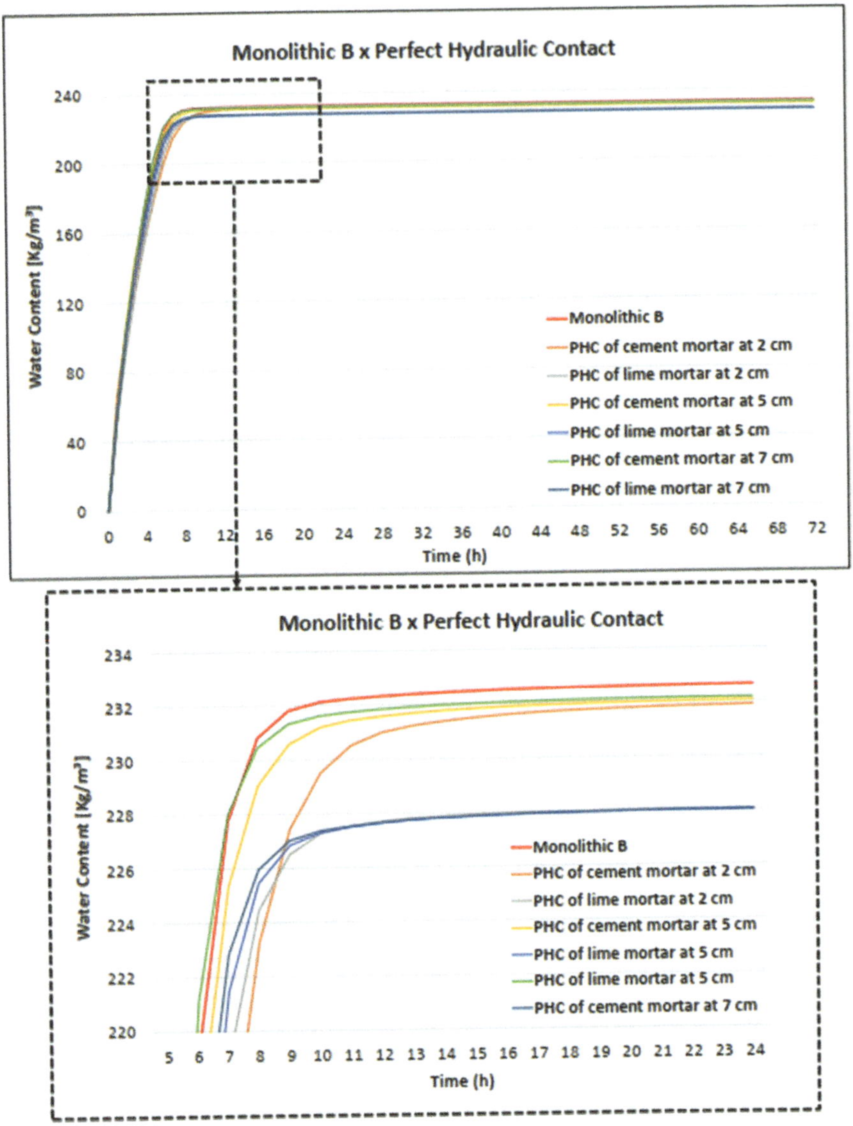

Fig. 14 Water content graph along brick B after 72 h of simulated transport for perfect hydraulic contact (cement × lime)

earlier, this drop point is characterized as the transition of moisture transport after the interface.

To understand the calculated result for the hydric resistance values, it is important to understand how the program uses material properties to interpret moisture transport. When the surface is completely wet (considering the presence of rain) the program calculates the liquid transport coefficient for suction using:

$$D_{ws}(w) = 3.8\left(\frac{A}{w_{sat}}\right)^2 1000^{\left(\frac{w}{w_{sat}}-1\right)} \tag{33}$$

where, D_{ws} is the liquid transport coefficient for suction (m^2/s), A is the water absorption coefficient ($kg/(m^2\sqrt{s})$), w_{sat} is the moisture content when the material is saturated (kg/m^3) and w is the moisture content (kg/m^3).

In Eq. (33) it is possible to identify the relationship between the moisture content and the water absorption coefficient for the liquid transport coefficient in the suction. Freitas et al. (1991) concludes in his study that the material permeability increases with the increase of the water absorption coefficient and that this coefficient causes an increase in the height reached at the wet front on the surface of the material.

Table 5 shows the values obtained for the hydric resistance of the simulated samples, using the same methodology as Azevedo (2019). The resistances values for brick A differ from the results obtained in the experimental test. This divergence may be related to the perfect hydraulic contact type interpreted by the calculation program that does not consider the hydric resistance caused by the discontinuity at the interface.

For brick A with cement mortar, a higher hydric resistance was identified for the sample with interface at 2 cm. Although the resistance values were very similar for the samples with interface at 5 cm and 7 cm, the lowest hydric resistance was presented for the sample with interface at 5 cm. For brick A with lime mortar, a higher hydric resistance was identified for the sample with 5 cm interface and the lowest hydric resistance was presented for the sample with interface at 2 cm.

The results for brick A samples with cement mortar proved to be consistent, considering that cement has a higher water absorption coefficient and liquid transport coefficient for suction than brick A. Therefore, the resistance is higher for interface located at 2 cm, because the liquid has to be transported through a thicker brick A layer to the top face.

The results for the brick A samples with lime mortar proved to be more challenging. Although lime has a lower water absorption coefficient and liquid transport in suction than cement mortar, but still higher than brick A's coefficients, it has a lower resistance for the sample with interface at 2 cm and higher for the sample with interface at 5 cm. The explanation may be related to the low water storage capacity (water content) for lime. Therefore, for the layer close to a higher water content, the resistance will be lower due to a higher water absorption by the second layer of brick A due to the lime layer saturation. For the interfaces at 5 cm and 7 cm, the saturated wet front does not approach the interface. The resistance is higher for the interface at 5 cm because it has a thicker brick A layer after the interface for moisture transport to the top.

For brick B samples with cement mortar and samples with lime mortar, higher water resistance was identified for samples with interface at 2 cm and lower resistance for samples with interface at 7 cm. The resistance results for brick B were consistent compared to the results obtained in the experimental test.

Table 5 Values of hydric resistance for the perfect hydraulic contact samples

Material	Sample/(interface type)	Hygric resistance (Azevedo 2019) kg/m^2s	Hygric resistance kg/m^3h
Red brick type "A"	Hydraulic contact (cement mortar at 2 cm)	9.7×10^{-5}	0.1421
	Hydraulic contact (cement mortar at 5 cm)	7.1×10^{-5}	0.0869
	Hydraulic contact (cement mortar at 7 cm)	4.2×10^{-5}	0.0871
	Hydraulic contact (lime mortar at 2 cm)	7.9×10^{-5}	0.1505
	Hydraulic contact (lime mortar at 5 cm)	5.8×10^{-5}	2.1055
	Hydraulic contact (lime mortar at 7 cm)	5.4×10^{-5}	1.0382
Red brick type "B"	Hydraulic contact (cement mortar at 2 cm)	7.9×10^{-5}	1.4384
	Hydraulic contact (cement mortar at 5 cm)	4.8×10^{-5}	0.4932
	Hydraulic contact (cement mortar at 7 cm)	3.2×10^{-5}	0.0274
	Hydraulic contact (lime mortar at 2 cm)	7.9×10^{-5}	2.4471
	Hydraulic contact (lime mortar at 5 cm)	4.2×10^{-5}	1.7953
	Hydraulic contact (lime mortar at 7 cm)	4.2×10^{-5}	1.2865

The higher resistance results in the interfaces located at lower heights are explained by the lower water absorption coefficient and liquid transport coefficient values of the mortars compared to the coefficient values for brick B. Thus, the mortar layer after the interface acts as a water barrier in transporting water through the second brick B layer. However, the moisture transport does not behave uniformly throughout the sample surface during the entire simulation period on account of the material interface influence and this influence will be shown in Figs. 15, 16, 17, 18, 19 and 20.

For an open scale (Fig. 15) it can be seen that the water content transport on the surface of the samples with brick A is similar for all samples when a low water content is transported on the wet front. Compared to the monolithic sample, there is a delay in moisture transport when transported across the interface due to the interface material's porosity being smaller than the porosity of brick A. For this condition the influence of the material interface on vapor water transport is verified. This condition is not verified when analyzing the transport of high water contents (Fig. 16).

Figure 16 indicates faster water transport across the interface for the samples with cement and lime compared to the monolithic sample A. The sample with cement

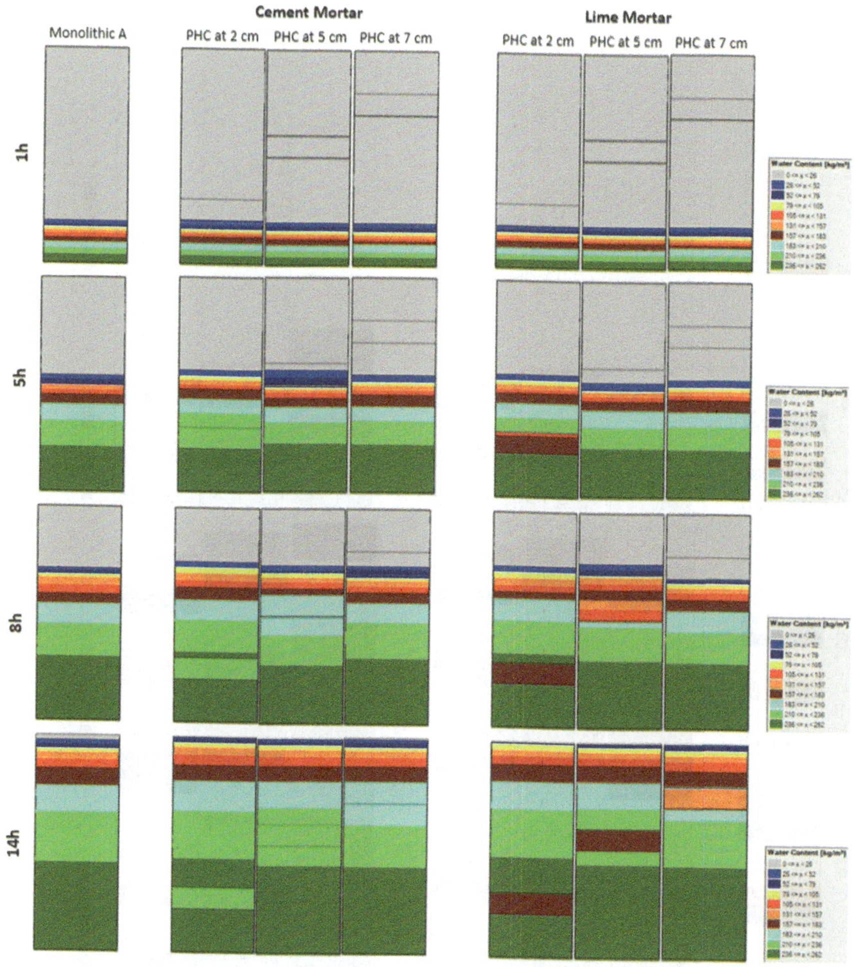

Fig. 15 Effect of perfect hydraulic contact on moisture surface progression after cement mortar and lime mortar interface at different simulation times for ceramic block A

interface at 2 cm shows slower water transport compared to the samples with interface at 5 cm and 7 cm. Although the interface at 2 cm is closer to a higher water content, this water is transported with greater resistance through brick A which is less permeable than concrete and has a lower liquid transport coefficient in suction. The water transport for the samples with interface at 5 cm becomes faster in relation to the samples with interface at 7 cm because it is closer to a higher water content.

The sample with lime interface at 2 cm shows faster water transport compared to the samples with interface at 5 cm and 7 cm. This result is directly related to the interface position height with access to the higher water content. The lime layer when it contains a high water content allows a faster transport through the second layer of brick A.

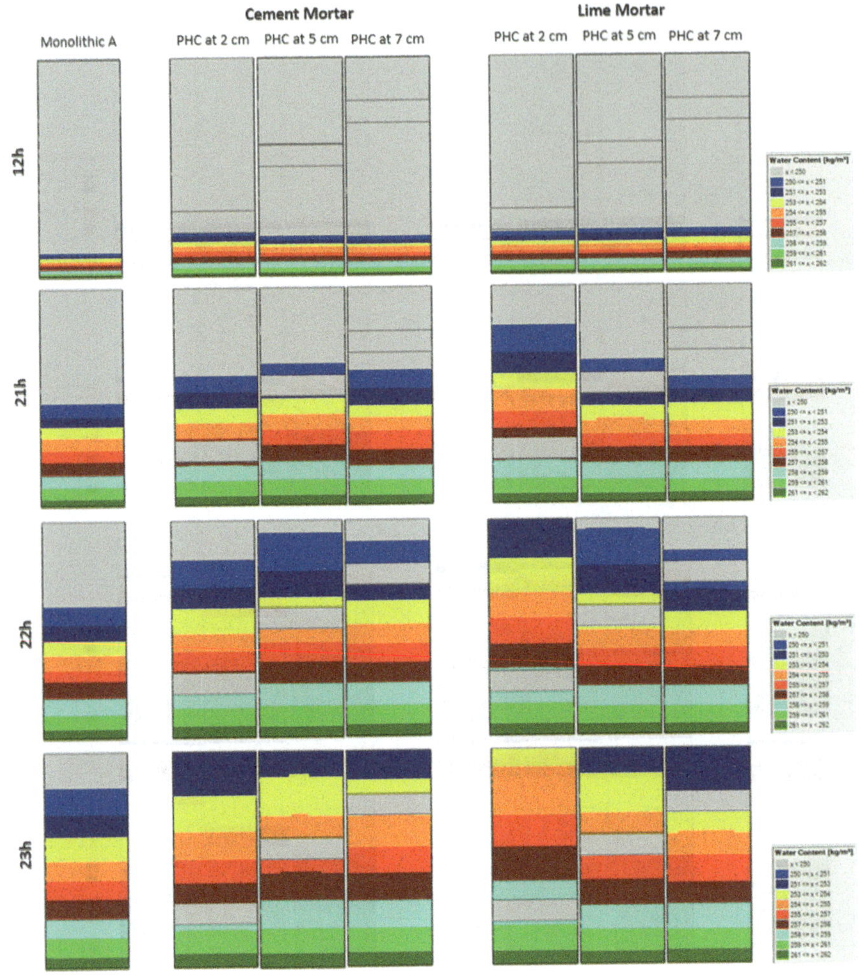

Fig. 16 Effect of perfect hydraulic contact on moisture surface progression (high water content) after cement mortar and lime mortar interface at different simulation times for ceramic block A

For the influence of the interface in brick A, one can conclude that the resistance to water transport has more influence on brick A due to its higher impermeability compared to mortars and will depend on the interface location with access to higher water content.

Figure 17 shows the water content on the surface of the samples with perfect hydraulic contact after 72 h of water transport simulation, compared to the monolithic sample A. For the samples with interface positioned at 2 cm with greater access to the saturated wet front, there is an increase in surface moisture content on the surface compared to the monolithic sample A. For the samples with interfaces far from the

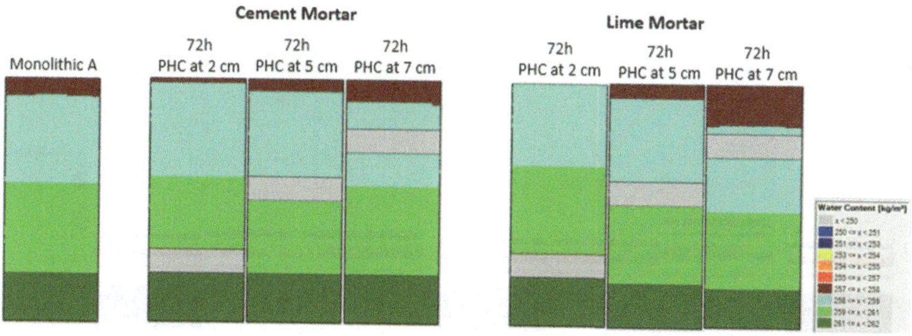

Fig. 17 Water content along the sample with brick A after 72 h of simulated transport for hydraulic contact

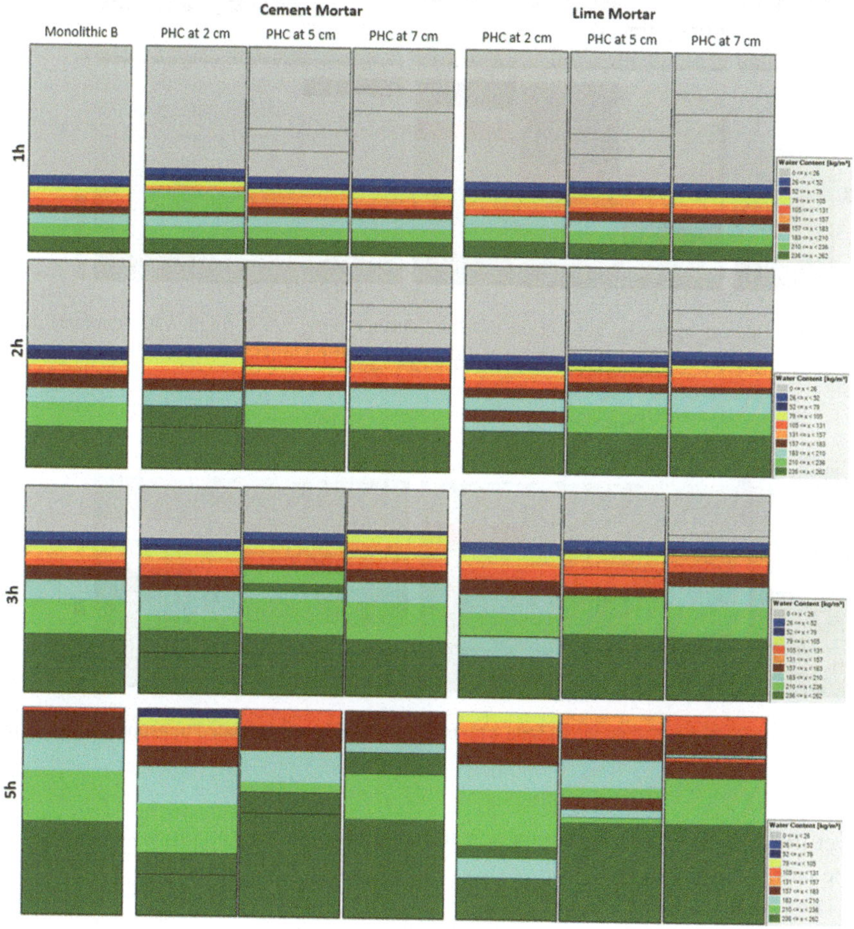

Fig. 18 Effect of hydraulic contact on moisture surface progression after cement mortar and lime mortar interface at different simulation times for ceramic block B

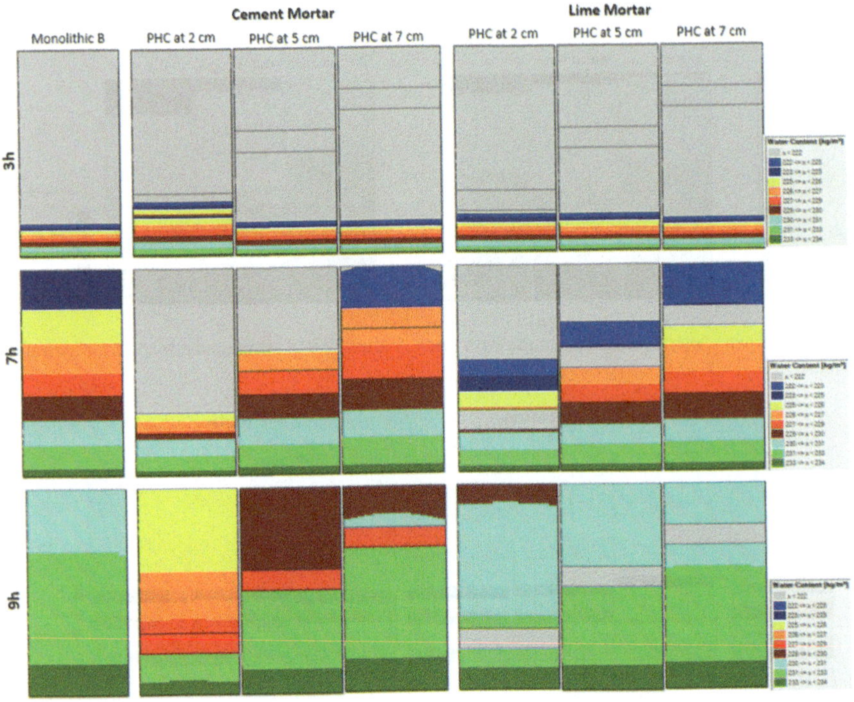

Fig. 19 Effect of hydraulic contact on moisture surface progression (high water content) after cement mortar and lime mortar interface at different simulation times for ceramic block B

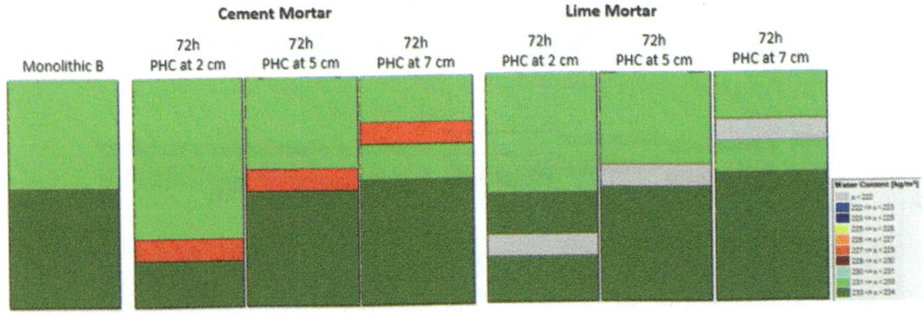

Fig. 20 Water content along the sample with brick B after 72 h of simulated transport for hydraulic contact

wet front, there is a visible reduction in the surface moisture content compared to the monolithic sample.

For an open scale (Fig. 18), it can be seen that the transport of the water content on the surface of the samples with brick B is similar for all samples when a low water content is transported on the wet front. Compared to the monolithic sample, there is also a small delay in moisture transport when transported across the interface due to the porosity of the interface material being smaller than the porosity of brick B. For this condition the material interface's influence on water vapor transport is verified. This condition is also verified when analyzing the transport of high water contents (Fig. 19). For brick B, this retardance is explained by the resistance to water transport through the interface of cement and lime mortar that are more impermeable than brick B (Table 6).

Therefore, the samples with interfaces located at 2 cm (Fig. 19) have higher resistance to water transport and the samples with interfaces located at 7 cm have lower resistance to water transport. Compared to the monolithic sample it is possible to observe the influence of the interface on water transport. Comparing the samples with cement mortar and lime mortar, the samples with lime mortar cause a lower

Table 6 Comparison of the velocity of moisture surface progression of different hydraulic interfaces (cement mortar and lime mortar), for brick A

Contact type PHC—BRICK A	Cement mortar			Lime mortar		
Interface level (cm)	2	5	7	2	5	7
Measured distance after the interface transition (cm)	8	5	3	8	5	3
Measured time (h)	13	9	6	13	9	6
Velocity (cm/h) of progression after the interface transition (cm/h)—water content	0.62	0.56	0.50	0.73	0.56	0.50
Measured time (h)	11	2	1	9	2	1
Velocity (cm/h) of progression after the interface transition (cm/h)—high water content	0.73	2.50	3.00	0.89	2.50	3.00
Velocity of progression after the interface transition (m/s) (Azevedo 2019)	2.3×10^{-7}	$2.9–1.9 \times 10^{-7}$	3.2×10^{-7}	4.8×10^{-7}	6.3×10^{-7}	3.8×10^{-7}

moisture transport resistance after the interface, due to the quick lime layer saturation which then facilitates the transport to the brick B layer. It can be concluded that the water transport resistance is higher for interfaces located at lower heights of the sample base, due to the higher impermeability of the mortars compared to brick B.

Figure 20 shows the water content on the surface of the samples with perfect hydraulic contact after 72 h of water transport simulation, compared to the monolithic sample B. For the samples with interface positioned at 2 cm with greater access to the saturated wet front, there is a reduction in wet front across the lime interface compared to the monolithic sample. For the cement interface, the wet front does not go beyond the cement layer. For the samples with interfaces far from the saturated wet front (interface at 7 cm), there is an increase in the height reached by the wet front compared to the monolithic sample.

Azevedo (2019) compared the progression of the wetting surface of different hydraulic interfaces (cement lime and mortar lime) by calculating the velocity at which the wet front overcomes the interface and reaches the top face of the sample.

Similarly, Tables 7 and 8 presents the velocities obtained from the simulation times for the moisture surface progression of the samples after the transition from the interface to the top face. As the results presented by WUFI-2D are only in hour interval, the image analysis of the moisture surface progression is necessary to better understand the transport.

For the samples with brick A, the transport of the low water content was faster in the samples positioned at 2 cm. The transport through the lime was faster compared to the cement. For the samples with interfaces at 5 and 7 cm the velocity values were the same, but for the same time, through the surface of the samples with lime, higher water contents were transported. A higher transport velocity can then be considered for the samples with perfect hydraulic contact with the lime mortar.

For the samples with brick A, the transport of high water content was slow for the samples positioned at 2 cm and greater for the interfaces positioned at greater heights (5 cm and 7 cm).

The velocity values found for the samples with brick A were consistent with the values presented in the experimental test for the samples with cement mortar, but diverged in the values found for the lime mortars.

For samples with brick B, the velocity results of the surface moisture progression in the samples showed similar behavior to those obtained from samples with brick A. Although, in comparison with the values obtained in the experimental test, greater divergences were presented. However, in comparison with the values obtained in the experimental test, greater divergences were presented.

However, it can be concluded that this methodology for analyzing the velocity of moisture progression at the interface is not clear, considering that the interval between the analyses is 1 h and the water content carried on the surfaces are different in the samples. The best way to analyze the surface moisture progression is by analyzing the transported water content through the simulation images that also allow access to the value of the water content at any point on the surface. Analyzing the images presented above (Figs. 15, 16, 17, 18, 19 and 20), it can be concluded that the water progression on the surface of the samples is faster in the samples with lime mortar

Table 7 Comparison of the velocity of moisture surface progression of different hydraulic interfaces (cement mortar and lime mortar), for brick B

Contact type PHC–BRICK B	Cement mortar			Lime mortar		
Interface level (cm)	2	5	7	2	5	7
Measured distance after the interface transition (cm)	8	5	3	8	5	3
Measured time (h)	4	3	2	4	3	2
Velocity (cm/h) of progression after the interface transition (cm/h)—water content	2.00	1.67	1.50	2.00	1.67	1.50
Measured time (h)	6 h	2 h	2 h	4 h	2 h	2 h
Velocity (cm/h) of progression after the interface transition (cm/h)—high water content	1.33	2.50	3.00	2.00	2.50	3.00
Velocity of progression after the interface transition (m/s) (Azevedo 2019)	3.5–2.3 × 10^{-7}	5.6–2.9 × 10^{-7}	1.8 × 10^{-7}	3.5 × 10^{-7}	6.3 × 10^{-7}	3.8 × 10^{-7}

Table 8 Values of hydric resistance for air space interface between brick layers

Material	Sample/(interface type)	Hydric resistance (Azevedo 2019) kg/m^2s	Hydric resistance kg/m^3h
Red brick type "A"	Air space 2 mm at 2 cm	0.8×10^{-5}	0.3570
	Air space 5 mm at 2 cm	0.4×10^{-5}	0.3469
	Air space 2 mm at 5 cm	0.9×10^{-5}	0.2154
	Air space 5 mm at 5 cm	0.8×10^{-5}	0.3800
	Air space 2 mm at 7 cm	0.9×10^{-5}	0.1989
	Air space 5 mm at 7 cm	0.4×10^{-5}	0.2944
Red brick type "B"	Air space 2 mm at 2 cm	0.7×10^{-5}	0.1569
	Air space 5 mm at 2 cm	0.8×10^{-5}	0.1450
	Air space 2 mm at 5 cm	1.0×10^{-5}	0.3325
	Air space 5 mm at 5 cm	0.7×10^{-5}	0.9219
	Air space 2 mm at 7 cm	0.7×10^{-5}	0.1313
	Air space 5 mm at 7 cm	0.9×10^{-5}	0.4144

compared to the samples with cement mortar. This result is consistent with the result presented in the experimental study.

Also, it can be concluded that the transport of low water content in the samples is slower than the transport of high water content. The progression of water on the surface of the samples is faster in the samples with brick B compared to the samples with brick A.

In resume:

- There is a reduction in the absorbed water content in the samples with cement mortar and lime mortar perfect hydraulic contact compared to the absorbed water content by the monolithic sample; This result was consistent with the Azevedo (2019) experimental result.
- No significant change was observed in the curves for samples with perfect hydraulic contact in comparison with the curve for monolithic samples. Therefore, not characterizing a discontinuity caused by the interface of the materials in the mortar samples, but considering the layer material properties as influencing the water transport resistance. This result diverges from the experimental result of Azevedo (2019).
- The samples with perfect hydraulic contact interface with cement mortar show higher absorption rates than the samples with lime mortar; This result diverged with the Azevedo (2019) experimental result. As well as for the previous divergence, the explanation for this result may be related to the type of perfect contact that the program calculates. In the experimental test the type of contact is imperfect, with a discontinuity ("layer") at the interface of unknown property caused by the brick layer absorbing particles from the fresh cement mortar layer, characterizing an imperfect hydraulic contact and attributing a higher resistance to water transport.
- Brick A proved to be more impermeable than cement and lime mortars, considering that brick A has a lower water absorption coefficient. This impermeability interfered in the water transport through the top layer of the samples, causing a slower absorption.
- The cement and lime mortars proved to be more impermeable than brick B, considering that brick B has a lower water absorption coefficient. This impermeability interfered in water transport through the interface of perfect hydraulic contact, causing a water barrier and slowing the water transport through the interface.
- The moisture progression on the brick B samples surface were faster than the moisture progression on the brick A samples surface. This result proved to be consistent with the experimental result; The velocity influence may be related to the higher permeability and surface area of brick B.
- The moisture progression on the surface of monolithic brick A sample is slower compared to the samples with perfect hydraulic contact interface due to higher mortar permeability compared to brick A permeability.
- It was found that the water storage function of the material in the joint will influence the wet front height on the surface of the most impermeable brick sample.

The less distant from the saturated wet front is the mortar interface with low water storage capacity, the higher the wet front on the surface of the sample.
- The material layer less permeable than the brick in the joint will reduce the water absorption in the sample when positioned at a shorter distance from the saturated wet front.

4.2 Moisture Transport Across Air Space Interface—Absorption Process

In order to study the influence of an air space between material layers, moisture transport simulations were performed on samples with two configurations of air layers placed between the brick layers at different heights (as presented previously in Table 2): 2 mm air space positioned at 2, 5 and 7 cm from the base of the sample and 5 mm air space positioned at 2, 5 and 7 cm.

Figures 21 and 22 show the plot of the simulations after 72 h for the moisture transport (kg/m^3) per hour along the samples with an air space interface between brick A and brick B layers, comparing with the moisture transport behavior in the monolithic brick sample.

Figures 21 and 22 reveal a water resistance when the water content reaches the interface (at 2 cm, 5 cm and 7 cm height), reducing and almost stopping the water content transport across the interface. The results are similar for the samples with 2 mm air space and 5 mm air space, but by analyzing the simulation results, a higher water content transported through the samples with 2 mm air space was observed.

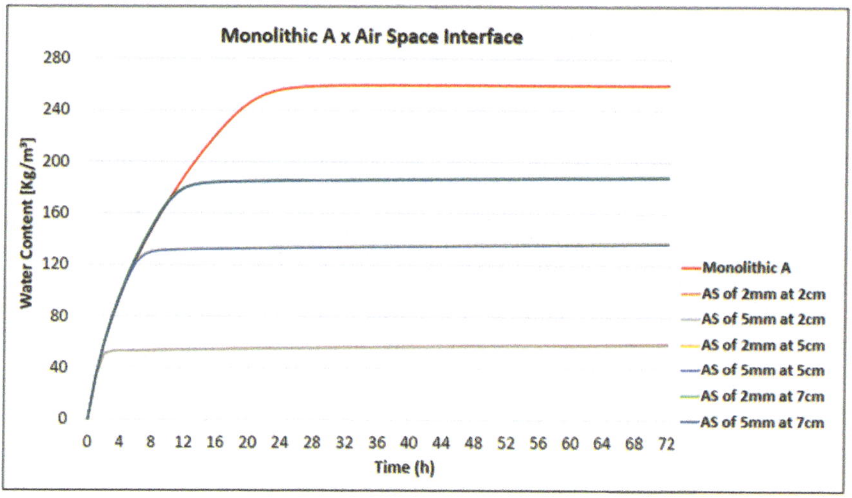

Fig. 21 Water content graph along brick A after 72 h of simulated transport for air space interface between brick A layers

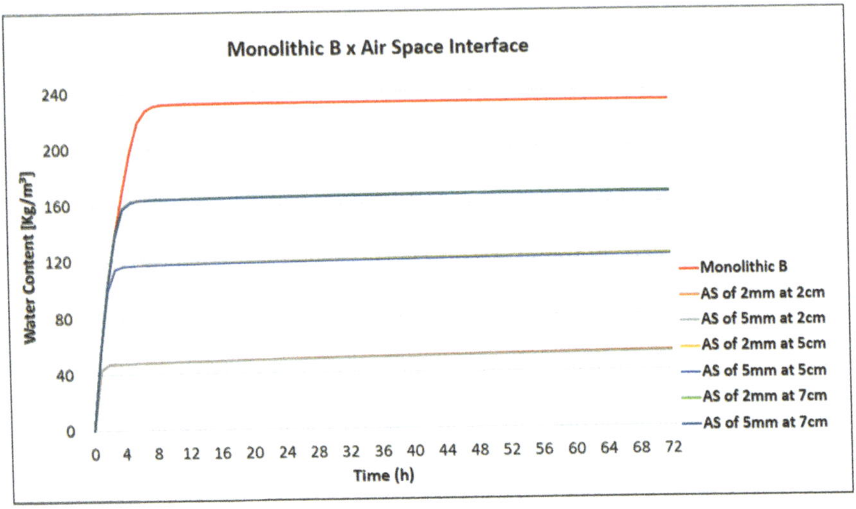

Fig. 22 Water content graph along brick A after 72 h of simulated transport for air space interface between brick B layers

Table 8 shows the values obtained for the hydric resistance of the simulated samples, using the same methodology as Azevedo (2019).

The presented results for hydric resistances at the air space interface were similar in samples with brick A and samples with brick B. For the samples with interface positioned at 2 cm, the samples with 2 mm air space interface showed higher resistances compared to the samples with 5 mm air space interface. For the samples with interface positioned at 5 cm and 7 cm, higher resistances were presented for the 5 mm air space interface. It is concluded that for 2 mm air layers the influence on water transport resistance is greater when the interface is close to the saturated wet front. For 5 mm air layers the influence is greater at interfaces positioned at 5 cm compared to samples with an interface positioned at 7 cm. According to the properties of the air layers, the 5 mm air layer has a higher water vapor resistance compared with the 2 mm air layer.

In the WUFI physics background, in vapor-tight assemblies, the high moisture capacity of these air layers may distort the moisture distribution, since the air absorbs too large a fraction of the moisture present in the assembly. In vapor-permeable assemblies this is less of an issue because the high moisture content accumulating in the air layer is not at the expense of the moisture in the other materials.

The hydric resistances for the simulated samples show a mostly different behavior than the resistance values found for the experimental samples.

In Fig. 23 brick A samples with interface positioned at 2 cm shows slower moisture transport through the 2 mm air space layer compared to the 5 mm air space layer. For the interfaces positioned at 5 m and 7 cm, the moisture transport is slower through the 5 mm air layer compared to the 2 mm air layer. In Fig. 24 brick B samples show the same moisture transport behavior as brick A samples. The results prove the consistency of the values found for the resistances presented in Table 7.

Figures 25 and 26 show the water content in the samples with air space interface at the end of the 72 h water absorption test simulation. In order to check the water content progression height on the surface after crossing the airspace interface, the water content values at different points on the surface of the samples at the end of the simulation were measured (Table 9).

In Fig. 25 it can be seen that a low moisture content passes the air space interface in the samples with brick A. An image with reduced water content scale was selected to evaluate the water content in the second brick A layer. Samples with 5 mm airspace interface showed a wet front with higher water content compared to samples with 2 mm air space interface.

In Fig. 26 it can be seen that significant water contents do not pass the air space interface in the samples with brick B. An image with reduced water content scale was selected to evaluate the water content in the second brick B layer. Samples with

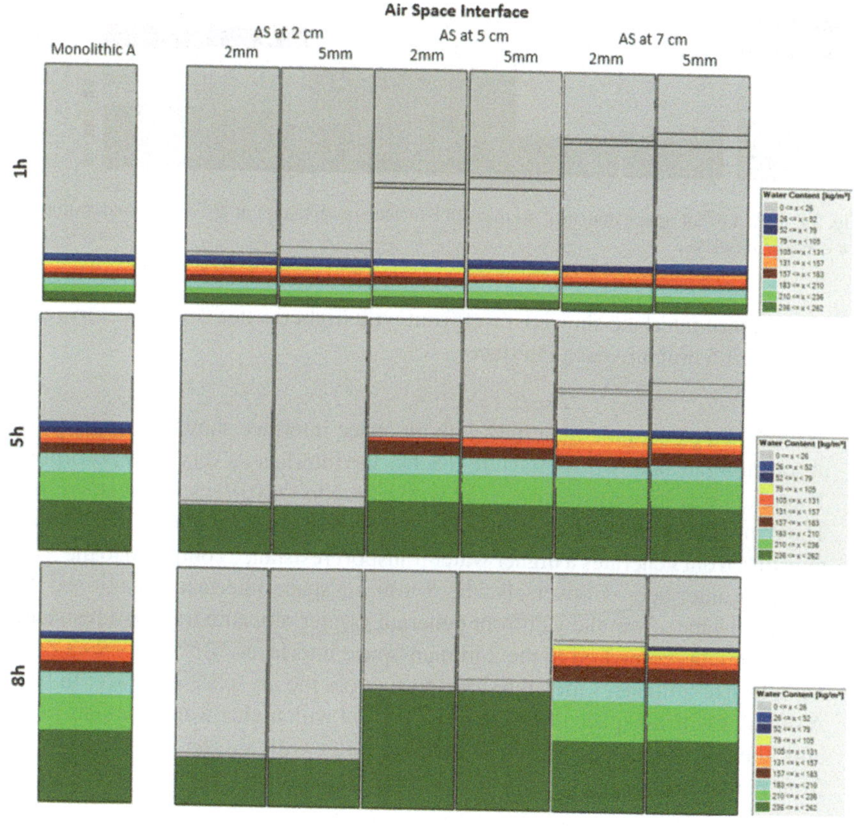

Fig. 23 Effect of air space interface on moisture surface progression at different simulation times for ceramic block A

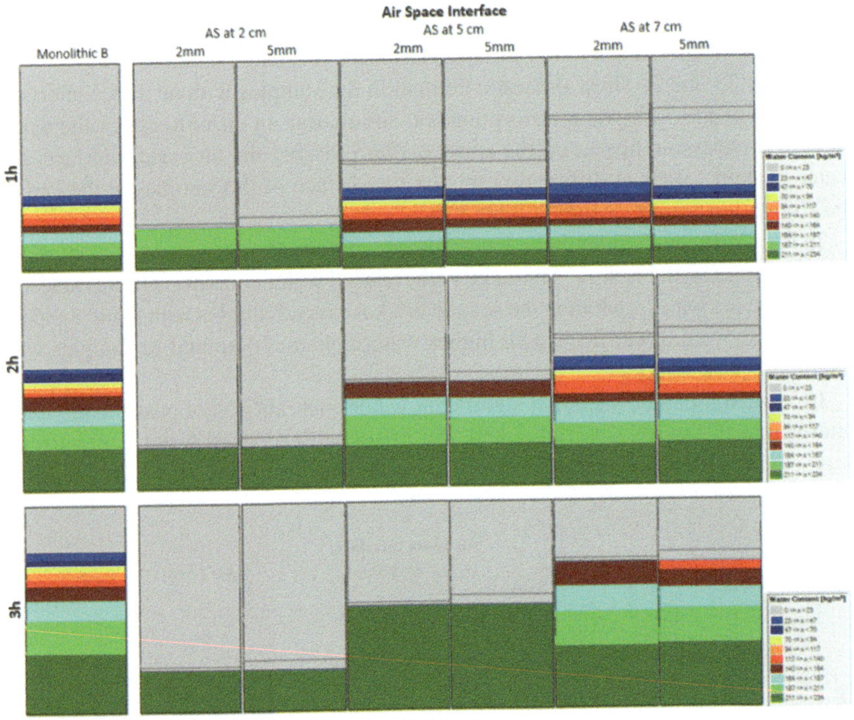

Fig. 24 Effect of air space interface on moisture surface progression at different simulation times for ceramic block B

2 mm air space interface showed a wet front with higher water content compared to samples with 5 mm air space interface.

In resume:

- The two tested ceramic materials with air space interface show a constant initial absorption rate and, when moisture reaches the interface, a very slow absorption; This result was shown to be consistent with Azevedo's (2019) experimental results.
- The 2 mm air space interfaces positioned at a shorter distance from the saturated moisture front generates a higher water transport resistance compared to the 5 mm air space interfaces. Conversely, the 5 mm air space interfaces positioned at a greater distance from the wet front generate greater moisture transport resistance in the samples compared to the 2 mm air space interfaces.
- Sample size influences moisture transport across the air space interface. In brick A samples with air space interface, the wet front with higher water contents raised after the 2 mm air layers. In the B-brick samples with air gap interface, the wet front with higher water contents raised after the 5 mm air layers.

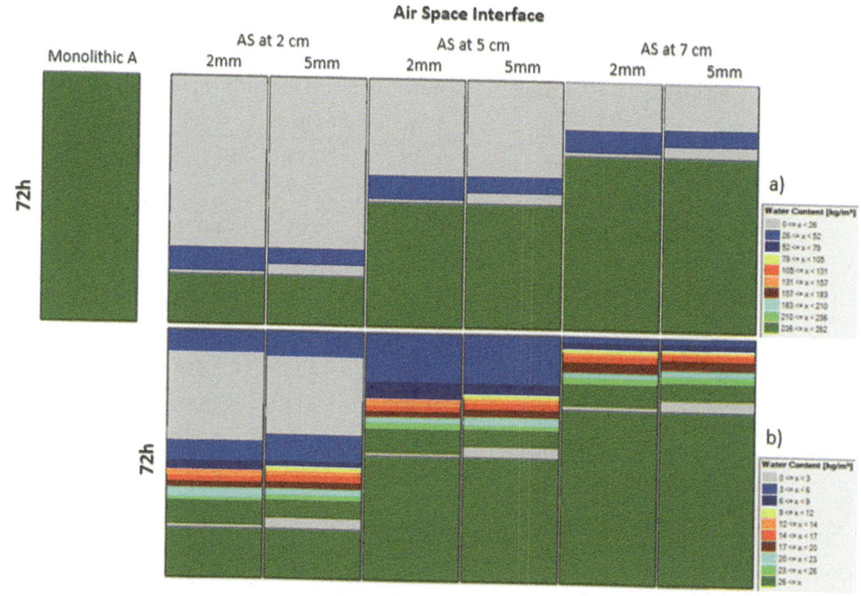

Fig. 25 Water content along the sample with brick A after 72 h of simulated transport for air space interface between brick A layers. **a** open scale, **b** reduced scale

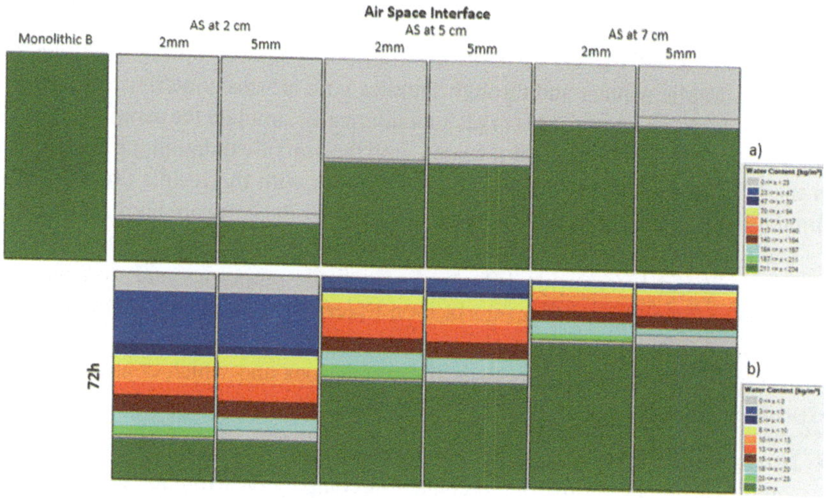

Fig. 26 Water content along the sample with brick A after 72 h of simulated transport for air space interface between brick B layers. **a** open scale, **b** reduced scale

Table 9 Water content values, at different points on the surface, for air space interface between brick layers

	Interface	Air space	Water content at the surface points (kg/m^3)			
			3 cm	6 cm	9 cm	10 cm
Brick A	2 cm	2 mm	27.75	2.672	–	–
		5 mm	27.99	2.672	–	–
	5 cm	2 mm	–	27.95	4.071	–
		5 mm	–	28.36	4.203	–
	7 cm	2 mm	–	–	16.52	4.276
		5 mm	–	–	17.47	4.316
Brick B	2 cm	2 mm	18.91	9.474	2.558	–
		5 mm	18.46	8.611	2.556	–
	5 cm	2 mm	–	19.35	9.357	2.774
		5 mm	–	18.92	8.772	2.687
	7 cm	2 mm	–	–	13.12	3.111
		5 mm	–	–	12.68	3.052

5 Conclusions

The development of this study made it possible to verify with the numerical calculation program WUFI-2D, the moisture transport behavior through samples with perfect hydraulic contact and through samples with air space interface.

In this work, an attempt was made to numerically simulate the samples studied in Azevedo's (2019) experimental research with the aim of validating the results found in the laboratory and resorting to further analysis with the results obtained in the simulation through the resources made available by the software application.

The results obtained allowed to develop a thorough analysis of water transport through material layers with different hygroscopic properties and to analyze the influence of the material layer interface on water transport during water absorption process in multilayer components.

The most significant results from the simulations and which contributed to the experimental study validation are the following:

- No significant change was observed in the water content transport curves for samples with perfect hydraulic contact in comparison with the curve for monolithic samples. Therefore, not characterizing a discontinuity caused by the interface of the materials in the mortar samples, but considering the layer material properties as influencing the water transport resistance. This result diverges from the experimental result of Azevedo (2019).
- The samples with perfect hydraulic contact interface with cement mortar show higher absorption rates than the samples with lime mortar; This result diverged

with the Azevedo (2019) experimental result. As well as for the previous divergence, the explanation for this result may be related to the type of perfect contact that the program calculates. In the experimental test the type of contact is imperfect, with a discontinuity ("layer") at the interface of unknown property caused by the brick layer absorbing particles from the fresh cement mortar layer, characterizing an imperfect hydraulic contact and attributing a higher resistance to water transport.

- With the results obtained in the water absorption test simulation using samples with perfect hydraulic contact, it could be concluded that in a brick and mortar masonry composed with a brick more impermeable than the mortar, the mortar layers with access to a significant moisture content will raise the moisture front on the surface of the component. In the case where the mortar of the composite is more impermeable than the brick, the mortar layer will act as a moisture barrier in the moisture transport.
- The material layer less permeable than the brick in the joint will reduce the water absorption in the sample when positioned at a shorter distance from the saturated wet front.
- It was found that the water storage function of the material in the joint will influence the wet front height on the surface of the most impermeable brick sample. The shorter the distance from the saturated wet front is the mortar interface with low water storage capacity, the higher the wet front on the surface of the sample, compared to a monolithic sample.
- The thinner air layer causes a higher water resistance in water transport through the air space interface.
- The brick permeability will influence the resistance caused by the air space interface in water transport in the drying process. The more permeable the second brick layer is, the smaller the effect caused by the water vapor diffusion resistance of the air layer.
- Although the software physics neglects the discontinuity effect caused by the interface between the material layers, the program proved to be helpful for studies of effective assemblies and solutions for moisture transport.

References

Azevedo AC (2019) Interface influence on moisture transport in building components. PhD thesis, Faculdade de Engenharia da Universidade do Porto, Portugal

Bear J, Bachmat Y (2012) Introduction to modeling of transport phenomena in porous media, vol 4. Springer Science & Business Media

Belarbi R, Qin M, Aït-Mokhtar A (2006) Modeling of moisture transport in porous building materials by gravimetric sorption-desportion tests, p 15

Benzaama MH, Rajaoarisoa LH, Ajib B, Lecoeuche S (2020) A data-driven methodology to predict thermal behavior of residential buildings using piecewise linear models. J Build Eng 32:101523

Brocken HJP (1998) Moisture transport in Brick Masonry: the grey area between bricks, PhD dissertation, Department Built Environment, Technische Universiteit Eindhoven

Chang SJ, Yoo J, Wi S, Kim S (2020) Numerical analysis on the hygrothermal behavior of building envelope according to CLT wall assembly considering the hygrothermal-environmental zone in Korea. Environ Res 191:110198
Crawley DB et al (2001) EnergyPlus: creating a new-generation building energy simulation program. Energy Build 33(4):319–331
Darcy H (1856) Les Fountaines Publiques de ville de Dijon. Dalmont Paris
Derluyn H, Janssen H, Carmeliet J (2011) Influence of the nature of interfaces on the capillary transport in layered materials. Construct Build Mater 25
Djedjig R, Bozonnet E, Belarbi R (2015) Analysis of thermal effects of vegetated envelopes: integration of a validated model in a building energy simulation program. Energy Build 86:93–103
Finite element analysis—theory and application with ANSYS. Miner Eng 12(8):992–993
Freitas V (1992) Transferência de humidade em paredes de edifícios – Análise do fenómeno de interface. PhD thesis, Faculdade de Engenharia da Universidade do Porto, Porto
Freitas V, Crausse P, E Abrantes V (1991) Moisture diffusion in thermal insulating materials. Seconde symposium on insulation materials: testing and applications, vol 2. ASTM STP 1116 a 400, Philadelphia, p 389
Freitas VP, Abrantes V, Crausse P (1996) Moisture migration in building walls—analysis of the interface phenomena. Build Environ 31:99–108
Freitas VP, Guimarães AS, Torres MI (2008) Humidade ascensional. FEUP edições, Porto
Gonçalves T (2007) Salt crystallization in plastered or rendered walls. PhD thesis, Instituto Superior Técnico e Laboratório Nacional de Engenharia Civil
Groot C, Gunneweg J (2010) The influence of materials characteristics and workmanship on rain penetration in historic fired clay brick masonry. Heron 55:141–154
Guimarães AS, Delgado JMPQ, Azevedo AC, de Freitas VP (2018) Interface influence on moisture transport in buildings. Construct Build Mater 162:480–488
Hens H (2002) Heat, air and moisture transfer in highly insulated building envelopes (HAMTIE). FaberMaunsell Ltd. https://www.ieaebc.org/Data/publications/EBC_Annex_24_tsr.pdf
Hens H (2007) Modeling the heat, air, and moisture response of building envelopes: what material properties are needed, how trustful are the predictions? JAI 4:1–11
Higashi K, Ito H, Oishi J (1963) Surface diffusion phenomena in gaseous diffusion. i. surface diffusion of pure gas. Nippon Genshiryoku Gakkaishi, vol 5
Iwamatsu M, Horii K (1996) Capillary condensation and adhesion of two wetter surfaces. J Colloida Interface Sci 182(2):400–406
Janssen H, Derluyn H, Carmeliet J (2012) Moisture transfer through mortar joints: a sharp-front analysis. Cement Concr Res 42:1105–1112
Janssens E (1998) Reliable control of interstitial condensation in lightweight roof systems. Ph.D. thesis, Catholic Universityof Leuven, Belgium
Karagiozis A (2001) Advanced hygrothermal modeling of building materials using MOISTURE-EXPERT 1.0. In: 35th international particleboard composite materials symposium, Pullman, Washington, USA
Karoglou M, Krokida MA, M.K., Maroulis, Z.B. (2007) A powerful simulator for moisture transfer in buildings. Build Environ 42(2):902–912
Kaviany M (1993) Principles of heat transfer in porous media. Springer-Verlag
Kerestecioglu A, Swami M, Gu L (1989) Combined heat and moisture transfer in building structures. ASME winter annual meeting, San Francisco, CA, 10-15 Dec 1989
Klein SA et al (2007) TRNSYS 16—a transient system simulation program. University of WisconsinMadison Solar Energy Laboratory, Madison, WI, USA
Kohonen R (1984) Transient analysis of thermal and moisture physical behavior of building constructions. Build Environ 19(1):1–11
Krus M (1996) Moisture transport and storage coefficients of porous mineral building materials—theoretical principles and new test methods. Ph.D. thesis, Fraunhofer IRB Verlag

Kuenzel MH (1995) Simultaneous heat and moisture transport in building components, one- and two-dimensionalcalculation using simple parameters. PhD thesis, Fraunhofer Institute in Bauphysics, IRB Verlag
Kumaran MK (1996) Heat, air and moisture transfer in insulated envelope Parts. Final report, vol. 3, task 3: material properties, international energy agency annex 24, Laboratorium Bouwfysica, K. U.-Leuven, Belgium, p 135
Künzel H, Kiesel K (1997) Calculation of heat and moisture transfer in exposed building components. Int J Heat Mass Transf 40(1):159–167
Langmans J, Nicolai A, Klein R, Roels S (2012) A quasi-steady state implementation of air convection in a transient heat and moisture building component model. Build Environ 58:208–218
Lewis WK (1921) The rate of drying of solid materials. J Ind Eng Chem Symp Drying, pp 427–432
Liang C, Chen X (1999) Exploration electromagnetic field theory. China Mining University Press', Xuzhou
Luikov AV (1966) Heat and mass transfer in capillary-porous bodies. Pergamon Press
Mason EA (1983) Gas transportin porous media: the dusty-gas model
Mendes N, Ridley I, Lamberts R, Philippi PC, Budag K (1999) UMIDUS: a PC program for the prediction of heat and moisture transfer in porous building elements. Building simulation conference-IBPSA, pp 277–283
Mendes N, Philippi PC, Lamberts R (2002) A new mathematical method to solve highly coupled equations of heat and mass transfer in porous media. Int J Heat Mass Transfer 45:509–518
Murray MC et al (2009) Live energy TRNSYS: TRNSYS simulation within Google Sketchup, p 8
Nakasone Y, Yoshimoto S, Stolarski TA (2006) Engineering analysis with ANSYS software. Butterworth-Heinemann, Amsterdam; Boston
Nicolai A, Grunewald J, Zhang JS (2007) Calculation of heat and moisture transfer in exposed building components. Salztransport und Phasenumwandlung - Modellierung und numerische Lösung im Simulationsprogramm Delphin 5, Bauphysik, pp 231–239
Pedersen CR (1992) Prediction of moisture transfer in building constructions. Build Environ 27(3):387–397
Pel L (1995) Moisture transport in porous building materials. PhD thesis, Einhoven University of Technology, The Netherlands
Philip JR, De Vries DA (1957) Moisture movement in porous media under temperature gradients. Trans Am Geophys Union 38:222–232
Preuss J (2015) Moisture balance. In: TRaNsient SYstem simulation program 18: multizone building modeling with Type56 and TRNBuild, TRNSYS, pp 191–195
Qiu X (2003) Moisture transport across interfaces between building materials. PhD thesis, Universidade Concordia, Montreal
Richards LA (1931) Capillary conduction of liquids through porous mediums. Physics 1:318–333
Rode C (1990) Combined heat and moisture transfer in building constructions. Ph.D. thesis, Thermal Insulation Laboratory, Technical University of Denmark
Rouchier S (2012) Hygrothermal performance assessment of damaged building materials. Université Claude Bernard - Lyon I, Materials
Ruisinger U, Kautsch P (2020) Comparison of hygrothermal 2D- and 3D-simulation results with measurements from a test house, E3S web conference, vol 172, p 08004
Schirmer R (1938) ZVDI, Beiheft Verfahrenstechnik, Nr. 6, S 170
Steeman M, Janssens A, Steeman HJ, Belleghem MV, Paepe MD (2010) On coupling 1D non isothermal heat and mass transfer in porous materials with a multizone building energy simulation model. Build Environ
Trechsel HR (ed) (1994) Manual on moisture control in buildings, ASTM M N L 18 American society for testing and materials, West Conshohocken, PA
Trechsel HR (2001) Moisture analysis and condensation control in building envelopes. 100 Barr Harbor Drive, PO Box C700, West Conshohocken, PA 19428-2959: ASTM International
Tuller M, Or D (2001) Hydraulic conductivity of variably saturated porous media: Film and corner flow in angular pore space. Water Resour Res 37(5):1257–1276

Verma SK et al (2022) Hygrothermal dynamics for developing energy-efficient buildings: building materials and ventilation system considerations. Energy Build 260:111932

Wang X et al (2011) The application of COMSOL multiphysics in direct current method forward modeling. Proc Earth Planet Sci 3:266–272

Whitaker S (1977) Simultaneous heat, mass, and momentum transfer in porous media: a theory of drying. Advances in heat transfer. Elsevier, pp 119–203

Zou Y (2020) Modelling of the dynamic effects in capillary pressure in coupling with deformation on the desiccation of porous materials. PhD thesis. L'école Centrale de Nantes

Numerical Simulation of Moisture Transport Along Ceramic Bricks—Drying Process

C. M. Araújo, A. C. Azevedo, and F. A. N. Silva

Abstract The drying process depends on internal and external factors such as temperature, relative humidity, the critical water content of the material, and the water transport properties in the liquid and vapor phases. However, the interface phenomena observed in multi-layered building components, as brick–mortar composites, or mortar-brick-insulation-mortar solutions, contribute to obtaining different values that result from the moisture transfer considering different materials/layers separately. The interface phenomena promote a hygric resistance which means that becomes a slowing moisture transport across the material interface. In this work a numerical study was carried out in order to analyse two different ceramic brick blocks with different interfaces, hydraulic contact interface, perfect contact interface, and air space), at different interface heights. The data used to run the simulations were taken from the wetting experiments on the samples; the corresponding moisture content profiles were measured using a gamma-ray spectrometer. Finally, the numerical results were compared with experimental values presented in the literature. The results showed an increase in the drying time constant for the materials with interface compared to monolithic materials. In addition, the farther away the interface is located from the base, the longer the drying time constant.

Keywords Ceramic brick · Numerical simulations · WUFI · Moisture transport · Interface · Drying process

C. M. Araújo · F. A. N. Silva
Civil and Engineering Department, Catholic University of Pernambuco, Recife, Pernambuco, Brazil
e-mail: camila.2020604019@unicap.br

F. A. N. Silva
e-mail: fernando.nogueira@unicap.br

A. C. Azevedo (✉)
Instituto Federal de Ciências de Educação e Tecnologia de Pernambuco (IFPE), Recife, Brazil
e-mail: antonio.costaazevedo@fe.up.pt

J. M. P. Q. Delgado (ed.), *Building Pathologies: Experimental Campaigns and Numerical Procedures*, Building Pathology and Rehabilitation 25,
https://doi.org/10.1007/978-3-031-17061-4_2

1 Introduction

1.1 Brief Review

In building construction, relative humidity is the cause of various pathologies that affect the durability and quality of buildings and interior spaces. Moisture not only degrades aesthetically, but also alters the characteristics and behavior of materials, mainly affecting their thermal properties.

Since the vast majority of construction materials used are porous materials, water accumulates in their empty spaces. Its porosity leads to the movement of water inside it through physical phenomena. It is through the absorption of water, of the heat flux, and mainly of the drying that there is a greater degradation of the materials.

Drying is a phenomenon of moisture transfer, in liquid or vapor form, from the interior of the material to the environment. It is necessary, for drying to occur, that there is sufficient energy to remove the water vapor from the material or to evaporate the water that is in the liquid state and, later, remove it from the interior of the porous material to the environment.

The study of the drying process for porous building materials has gained importance due to its importance for the durability of buildings and their maintenance. In this way, the drying kinetics commonly used in the food industry begins to be used more frequently for building materials. Drying kinetics models allow describing the drying process of materials.

Bednar (2002) and Van Belleghem et al. (2012) refer to three drying periods for porous materials. In a first period, the material adapts to the environmental conditions: the temperature of the material is reduced until the heat loss due to evaporation and the heat gain are equal. When the balance between the outgoing heat and the incoming heat is established, the second period is entered, in which the temperature and the drying flow of the material are constant and the moisture transfer from the interior of the material is equal to the humidity that goes out to the environment. In the third and last period the drying flow decreases, the object starts to get dry and the drying is conditioned by the amount of moisture that comes from the interior of the object to the surface. During this period, the temperature of the material begins to increase.

According to Krischer and Kroell (1963) three stages in the drying process for porous materials when it is saturated and placed in an environment with pre-defined temperature and relative humidity.

Currently, it is considered that the first phase of the drying phenomenon is controlled by liquid and/or vapor diffusion mechanisms. In the study of drying, thin-layer models are used, which presuppose a layer of material that is sufficiently thin so that the external conditions of the air remain constant on the material during drying. These models are, according to the literature, distributed in three categories (Panchariya et al. 2002; Kuitche et al. 2007), these being theoretical, semi-theoretical and empirical, and they also assume that the resistance to moisture flow is uniformly distributed throughout the interior of the isotropic homogeneous material.

Theoretical models consider internal resistance to moisture flow, while semi-theoretical and empirical models only reflect external resistance to moisture transfer between the material and the air. Theoretical models consist of solving the general drying equations, using an analytical or numerical method. Initially, some simplifications of analytical resolutions of Fick's second law were used. Crank (1975) gave analytic solutions for various products of regular geometry, with rectangular, cylindrical and spherical shapes. The combination of these solutions with the Arrhenius solution allowed the prediction of the drying process of some products, using numerical solutions of mass and heat transfer.

The semi-theoretical models generally arise from the simplification of Fick's second law solutions (which is a theoretical model) or from the modification of simplified and valid models within the limits of temperature, relative humidity, air velocity and threshold amount of water for the which were developed (Iglesias and Chirife 1983; Marinos-Kouris et al. 1996; Maroulis and Saravacos 2003). These models are easier to use, requiring little time to be used, compared to theoretical models. They are simplified and in some cases have the addition of empirical coefficients to improve the approximation of the drying curve to the experimental data. The difficulty in selecting a diffusion coefficient during drying makes it difficult to use this type of model. For reasons of simplification, a constant diffusion coefficient is chosen. In this way, the concept of the drying characteristic curve arises and the drying rate is expressed as a function of the water content of the material. However, this approximation does not represent a physical model of the drying phenomenon. Among the semi-theoretical models are the two-term model, the Henderson and Pabis model, the Page model, the modified Page model and the Newton model.

In building construction, several researchers (Mendes et al. 2003; Wilson et al. 1993; Qin et al. 2009; Pel 1993; Delgado et al. 2019) analyse the effect of perfect contact interfaces in multi-layered walls. However, related to models with imperfect contact, literature is less abundant. Only, as an example, Freitas et al. (1996) analyse the interface influence on the drying kinetics of samples of cellular concrete and red brick; Derdour et al. (2000) studied the effect of thickness, porosity and the drying conditions of several building materials on the drying time constant; Bednar (2002) studied the influence of the material size, insulation, and climate conditions on the moisture transport coefficient; and Karoglou et al. (2003) evaluated the effect of air temperature, air relative humidity, and air velocity, on the drying performance of building materials.

1.2 Drying Phases

When dealing with phenomena of high complexity, it is common to consider monolithic constructive elements since the existence of joints or layers contributes.

The drying process in a porous material can be defined as the removal of all water present in that material. The 1-D isothermal drying (one drying front) from the saturated state of the material, the specific interest of this study, is characterized by the following stages:

During the initial phase (Phase I), the material is saturated, with high moisture concentration at its surface, and the liquid water present in the porous matrix is moved to the upper surface through capillarity due to the high capillary pressure that exists at the surface. The moisture content in the material decreases through the surface by evaporation. The drying rate is controlled by environmental conditions such as wind speed, air temperature, air humidity, etc. (Hall and Hoff 2009). At this stage, there is a homogeneous moisture distribution and the drying rate is constant, with a linear decrease in moisture content over time and moisture transport within the material is faster than mass transfer to the atmosphere at the surface.

The second phase (Phase II) begins when the capillary flow is no longer sufficient to compensate for the amount of water evaporating. Thus, there is no longer an equilibrium between the liquid transported to the surface and the liquid that is being evaporated, causing a regression of the wet front into the interior of the material. The moisture transport mechanism to the surface of the material is by vapor diffusion above the wet front and by capillarity below the wet front. At this stage the drying rate declines, as the progressive decrease of the wet front increases the thickness of dry material that the vapor has to traverse to the surface. However, the drying process becomes dependent on the moisture transport properties of the material.

In the last phase (Phase III) the liquid discontinuity occurs below the moist zone. The beginning of this phase is unclear and is usually set as a virtual boundary (Selih et al. 1996). The liquid water transport in this phase is limited to the finest pores, which although visually the material looks dry, there are still small pores saturated or partially saturated. The main mechanism of moisture transport is vapor diffusion (Azenha 2009).

The mechanism implies that the drying is complex, multiple and interdependent (Mainguy et al. 2001). Figure 1 shows the representation of the three phases. The literature still presents few studies related to drying kinetics in porous materials. Freitas et al. (1996) studied the influence of the interface on the drying kinetics of cellular concrete and red brick samples. The effect caused by parameters such as thickness, porosity and the drying conditions were studied by Derdour et al. (2000).

2 WUFI-2D

In this work, the moisture performance simulation model WUFI 2D 4.3, which is a windows based program for the hygrothermal analysis of building envelope and components constructions, is selected. WUFI (Warme und Feuchte instationar—Transient Heat and Moisture) is a one-dimensional, two-dimensional or 3D model for heat and moisture transport developed by the Fraunhofer Institute in Building Physics (IBP) in Holzkirchen, Germany. Which can be used to assess the heat and moisture distributions for a wide range of building material classes and climatic conditions.

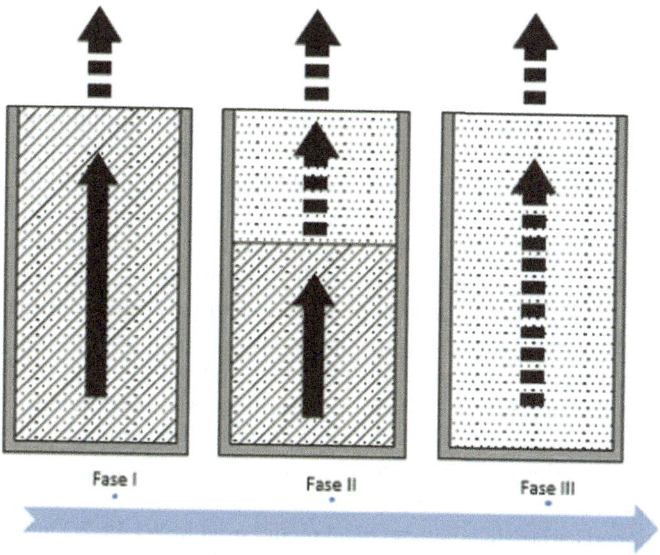

Fig. 1 One-directional drying of porous materials

The program requires only the standard material properties and allows the determination of moisture storage and liquid transport functions. It was validated through in situ and laboratory data, allowing a realistic simulation of the hygrothermal behavior of building elements, monolithic or multilayer, exposed to real climatic conditions. The two-dimensional analysis performed by the WUFI-2D program is based on the finite-volume method, allowing the modeling of building elements with complex geometries (Freitas et al. 2008).

The mathematical model used in WUFI was developed by Kuenzel (1995) based on Kiebl's theorem. In this model the non-steady heat and moisture transport processes in building components are described by the following coupled differential equations for heat transport and moisture transport:

$$\frac{\partial H}{\partial v}\frac{\partial v}{\partial t} = \frac{\partial}{\partial x}\left(\lambda\frac{\partial v}{\partial x}\right) + h_v\frac{\partial}{\partial x}\left(\frac{\delta}{\mu}\frac{\partial p}{\partial x}\right) \tag{1}$$

$$\rho_w\frac{\partial u}{\partial \varphi}\frac{\partial \varphi}{\partial t} = \frac{\partial}{\partial x}\left(\rho_w D_w\frac{\partial u}{\partial \varphi}\frac{\partial \varphi}{\partial x}\right) + \frac{\partial}{\partial x}\left(\frac{\delta}{\mu}\frac{\partial p}{\partial x}\right) \tag{2}$$

where, Dw [m^2/s] is the liquid transport coefficient, H [J/m^3] is the enthalpy of moist building material, hv [J/kg] is the evaporation enthalpy of water, p [Pa] is the water vapor partial pressure, u [m^3/m^3] is the water content, δ [kg/msPa] is the water vapor diffusion coefficient in air, v [°C] is the temperature, λ [W/mK] is the heat conductivity of moist material, μ [–] is the vapor diffusion resistance factor of dry material, ρw [kg/m^3] is the density of water and φ [–] is the relative humidity.

3 Numerical Simulation

In the previous chapter it was possible to understand the functionality of the WUFI-2D program that will be used in this study to simulate moisture transport in samples with different contact configuration between the materials. Characteristic information of the simulations performed in this study was also provided.

The simulations performed in this work aim to analyze the influence of the interface, during the drying process, in different types of contact—perfect hydraulic contact between mortar and brick and the air gap between brick layers—on moisture transport, through the verification of moisture content profiles obtained with the simulations performed. Also, it is intended to briefly compare the results of moisture content profiles obtained in the experimental study of Azevedo (2019) and the numerical simulations performed in WUFI-2D, in order to contribute to the experimental validation. For this purpose, the same material properties, initial conditions and boundary conditions of the experimental study were used in the execution of the simulations.

As mentioned earlier, the mathematical model used in WUFI-2D has as a potential conductor for the capillary moisture transport the relative humidity, which is continuous in multi-layered building components, i.e. continuous across material independent material boundaries. This is why the hydraulic contact configuration is considered as perfect, with no consideration of the second layer material penetrating into the first layer and the porous network discontinuity at the interface, thus treated as ideal between two active capillary materials for the contact condition of the configurations in the simulations performed.

Although the mechanisms of moisture transport in a single building material have been and continue to be widely studied, the hydraulic characteristics of the interface at different types of contacts between materials are still poorly understood and, for this reason, the simplified assumption of perfect contact is widely used in hygrothermal models. However, compared to a monolithic element, the multilayer element exhibits delayed liquid transport across the interface between the layer materials (Azevedo 2019).

The geometry and properties of the samples to be analyzed took into consideration two types of red ceramic bricks: Brick A and Brick B; two types of mortars: Cementitious and Lime; three configurations for the application of hydraulic joint for each type of brick and each type of mortar; and six configurations for air gap application between brick layers for each type of brick. The simulations of monolithic bricks samples are used as a basis for comparing the effects caused by the interface on the samples.

The configurations of the simulated models are presented in Figs. 2 and 3 and Table 1, the properties and initial conditions of the materials used in the simulation are presented in Tables 2 and 3, respectively.

For the drying process simulation, adiabatic systems were applied for all surfaces, except for the top face that was exposed to two different types of indoor climates:

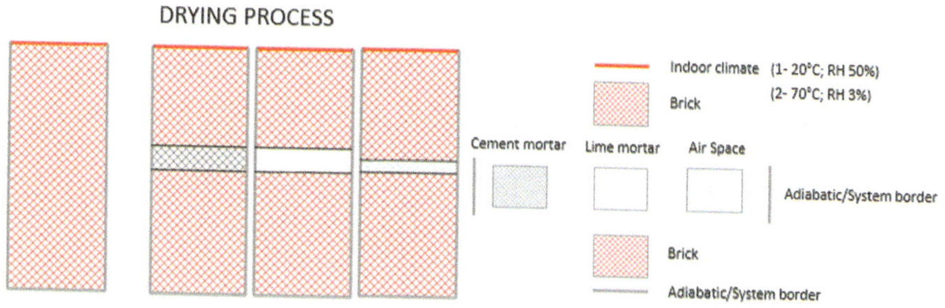

Fig. 2 Settings for the simulations processes

Environment 1—20 °C and RH 50% and Environment 2—70 °C and RH 3%. All materials had a saturated initial condition.

Currently the program has in the database new air layers "without additional moisture capacity". Their free saturation of 17 g/m^3 corresponds to the saturation moisture of air at 20 °C. However, the free saturation is always fixed at the value cited and does not vary with temperature, as it would in real air. The moisture contents and hygroscopic inertia of these air layers are at a realistic level, but can prove numerically challenging.

The simulations were performed for the same time period: 720 h—Environment 1 and 72 h—Environment 2. These parameters were used in the Azevedo (2019) experimental trial.

4 Results and Discussions

4.1 *Moisture Transport Across Hydraulic Contact—Drying Process*

In order to study the influence on the drying process of brick samples with perfect hydraulic contact (cement and lime mortar) at different heights, simulations were performed to compare the water transport behavior in the drying process for the monolithic samples and samples with perfect hydraulic contact aiming to find the influence of each contact configuration.

The samples used in the drying process were the same ones used in the absorption test. The samples had saturated initial condition for the materials and for the samples' base surface adiabatic system conditions were applied. The samples were exposed to two types of environments (Environment 1: T = 20 °C and 50% RH; Environment 2: T = 70 °C and 3% RH).

The drying process can occur in three phases: Phase I—When the material is saturated, with high moisture concentration on the material surface and the water transport occurs by capillarity. In this phase the drying rate is controlled by the environmental

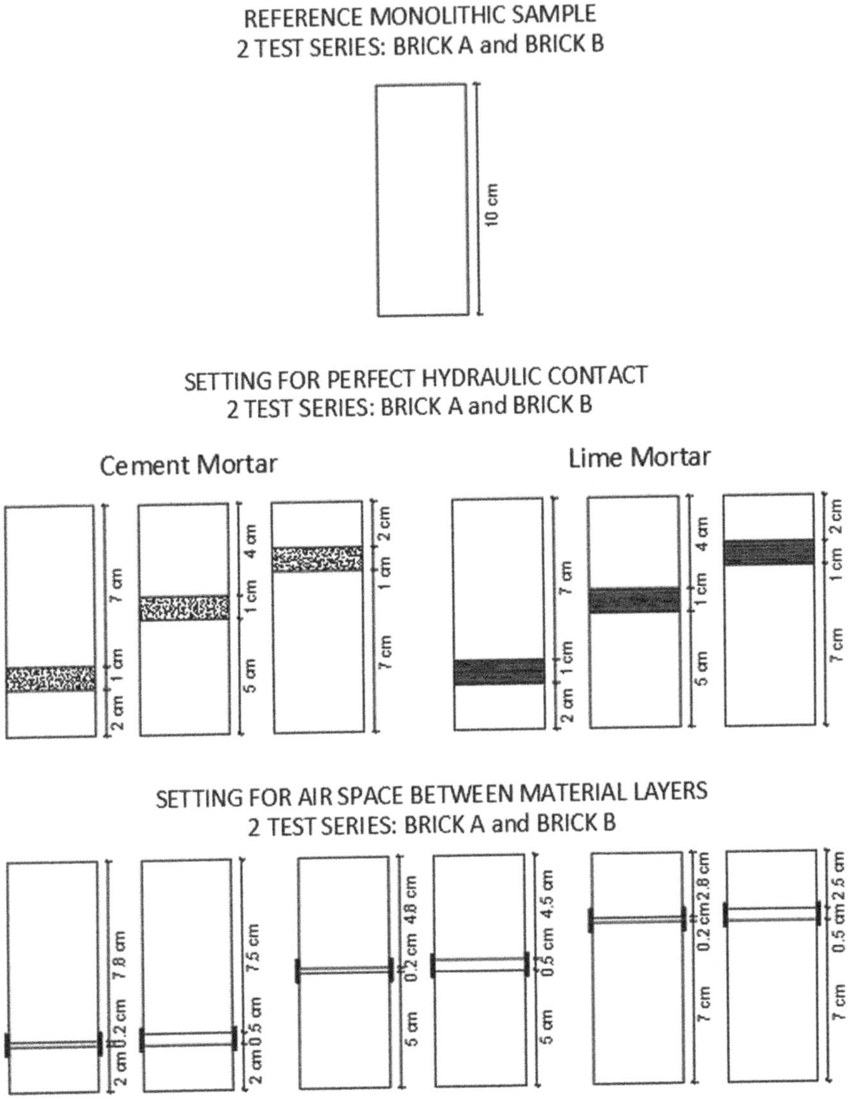

Fig. 3 Simulation setup: (**i**) monolithic block; (**ii**) a block with cement and lime mortar interface at 2 cm, 5 cm and 7 cm; (**iii**) an air space with 0.5 cm and 0.2 cm cavity at 2 cm, 5 cm and 7 cm

conditions, becoming constant due to the constant environmental conditions of the test. In Phase II there is no longer an equilibrium between the liquid transported to the surface and the liquid that is being evaporated, causing a regression of the wet front to the interior of the material and a reduction in the drying rate. In this phase the drying process becomes dependent on the moisture transport properties of the

Table 1 Settings for the simulations

Material	Dry environ 1 (20 °C, RH 50%); Dry environ 2 (70 °C, RH 3%)		
	Monolithic	Perfect hydraulic contact	Air space
Red brick type A 40 × 100 mm^2		Cement mortar at h = 20 mm	Cavity of 2 mm at h = 20 mm
		Cement mortar at h = 50 mm	Cavity of 5 mm at h = 20 mm
		Cement mortar at h = 70 mm	Cavity of 2 mm at h = 50 mm
		Lime mortar at h = 20 mm	Cavity of 5 mm at h = 50 mm
		Lime mortar at h = 50 mm	Cavity of 2 mm at h = 70 mm
		Lime mortar at h = 70 mm	Cavity of 5 mm at h = 70 mm
	1 sample	6 samples	6 samples
Red brick type B 50 × 100 mm^2		Cement mortar at h = 20 mm	Cavity of 2 mm at h = 20 mm
		Cement mortar at h = 50 mm	Cavity of 5 mm at h = 20 mm
		Cement mortar at h = 70 mm	Cavity of 2 mm at h = 50 mm
		Lime mortar at h = 20 mm	Cavity of 5 mm at h = 50 mm
		Lime mortar at h = 50 mm	Cavity of 2 mm at h = 70 mm
		Lime mortar at h = 70 mm	Cavity of 5 mm at h = 70 mm
	1 sample	6 samples	6 samples
Wetting	2 samples	12 samples	12 samples
Dry	2 samples	24 samples	12 samples
	4 simulations	36 simulations	24 simulations
Total	64 simulations		

material. Phase III is characterized by a slow, residual drying rate, close to the steady state condition and the main moisture transport mechanism is vapor diffusion.

The curves in Fig. 4 showed different drying trends for the brick A samples with cement mortar and the samples with lime mortar. In drying with RH 50% and T = 20 °C environment, both samples experienced an initial long linear water loss characterized by Phase I of the drying process. However, the samples with cement mortar initiated a faster Phase II compared to the samples with lime mortar, presenting a slow progressive drying behavior. This result is consistent with the result found

Table 2 Hygrothermal properties of materials

Properties	Materials/layers					
	Red brick type A	Red brick type B	Cement mortar	Lime mortar	Air space 2 mm	Air space 5 mm
Bulk density, ρ [kg/m^3]	1800	1600	1878	1810	12.9	12.9
Porosity, ε [m^3/m^3]	0.32	0.38	0.20	0.21	0.999	0.999
Specific heat capacity, C [J/kgK]	850	850	900	900	1000	1000
Thermal conductivity (Dry), λ. [W/mK]	0.40	0.38	0.786	0.799	0.0125	0.02778
Water vap. diff. resist. factor, μ [–]	33.1 (dry cup)	21.4 (dry cup)	23.90	14.25	5.59	3.224
Moisture Content, w80% [kg/m^3]	w80% = 15.51	w80% = 9.75	w80% = 38.16	w80% = 8.16	0.0136	0.0136
Moisture Content, wsat [kg/m^3] Free water saturation	wsat = 261.38	wsat = 233.07	wsat. = 228.59	Wsat. = 186.61	0.017	0.017
Water absorption coeff. [kg/m$^2\sqrt{s}$]	0.10	0.19	0.15	0.12	–	–

for the water absorption simulation in the lime samples which is faster compared to the monolithic brick A sample. Therefore, in the drying process, the transport phase by capillarity and vapor diffusion shows to be more favorable in the samples with perfect hydraulic contact with lime mortar, in comparison with the monolithic brick A sample, which has a lower liquid transport coefficient and a lower water vapor permeability.

However, for none of the perfect hydraulic contact configurations was observed a significant change in the curves compared to the curve of the monolithic sample, not characterizing a discontinuity caused by the materials interface in the samples with mortar. This result diverges from Azevedo's (2019) experimental result that found a moisture content discontinuity across the interface, indicating a capillary pressure difference across the interface. With this, it can be concluded that the assumption of perfect contact for samples with hydraulic contact is negligent in the numerical calculation.

Table 3 Initial conditions of materials/layers

Initial conditions			
	Temperature (°C)	Moisture content, w (kg/m^3)	Relative humidity, RH (%)
Red brick	20	w80% = 15.51	0.8
type A		wsat. = 261.38	1
Red brick	20	w80% = 9.75	0.8
type B		wsat. = 233.07	1
Cement mortar	20	w80% = 38.16	0.8
		wsat. = 228.59	1
Lime mortar	20	w80% = 8.16	0.8
		wsat. = 186.61	1
Air space	20	w80% = 0.0136	0.8
		wsat. = 0.017	–

Figure 5 shows the water content on the brick A samples' surface at the end of the drying process simulation. The samples with hydraulic contact have lower water content on the surface, and it can be concluded that they have a higher drying rate compared to the monolithic sample which is more impermeable and retains more water due to capillary pressure in the small pores.

Figure 5 also verifies that the interfaces in the samples with brick A located at a greater distance from the base present a lower moisture content progression to the surface, requiring more time in the drying process for these samples. In all samples, the second layer of brick A exposed to the ambient conditions present lower water contents, which characterizes a greater drying in the region. These results were proven for the Azevedo's (2019) experimental study.

The curves in Fig. 6 displayed similar drying tendency for the brick B samples with cement mortar and the samples with lime mortar. In both drying environments, both samples experienced a long initial linear water loss characterized by Phase I capillary transport of the drying process. The samples with cement mortar showed faster drying in Phase I and slower drying in Phase II compared to the samples with lime mortar and the monolithic sample. This result is coherent with the water absorption process simulation result, where the samples with mortar showed a faster water transport by capillary absorption, which makes the second drying phase slower by the water shortage and a water vapor diffusion process is initiated earlier compared to the monolithic sample and lime mortar sample.

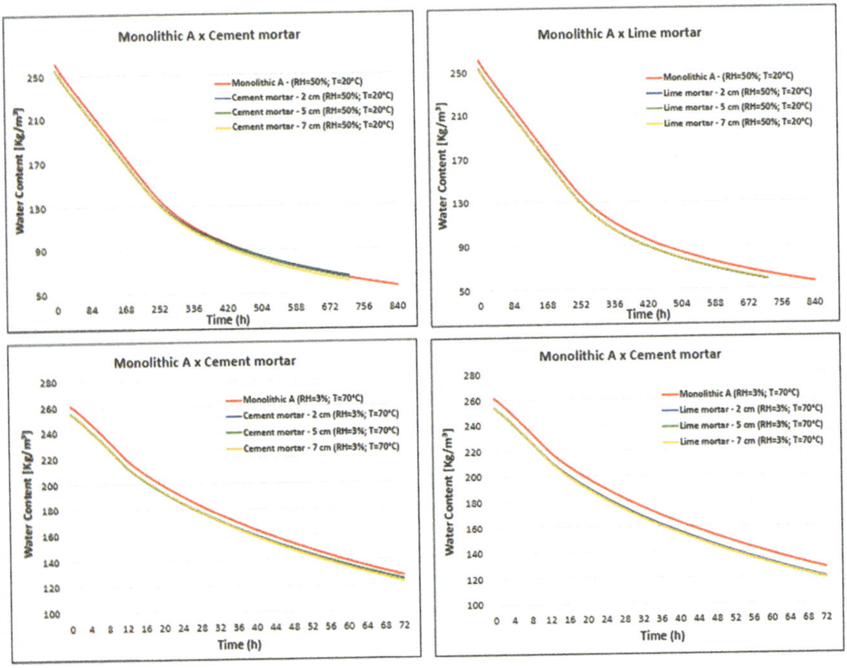

Fig. 4 Water content graph along brick A of simulated transport on the drying process for perfect hydraulic contact (cement × lime)

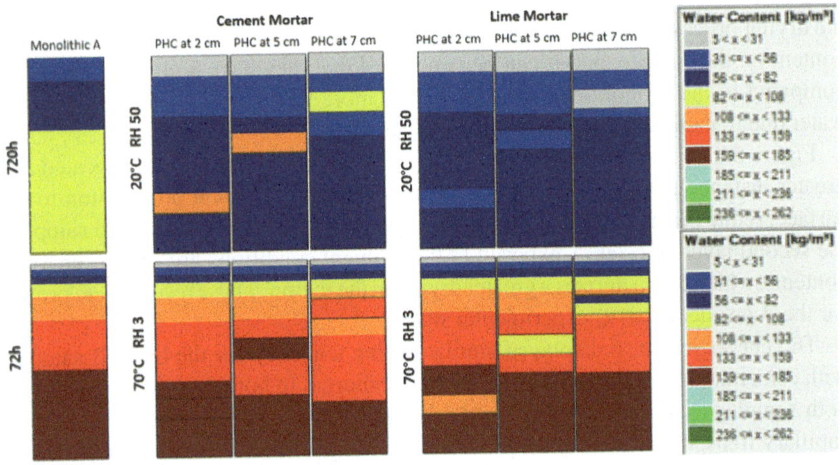

Fig. 5 Water content on the surface of samples with brick A at the end of simulated transport on the drying process for perfect hydraulic contact (cement × lime)

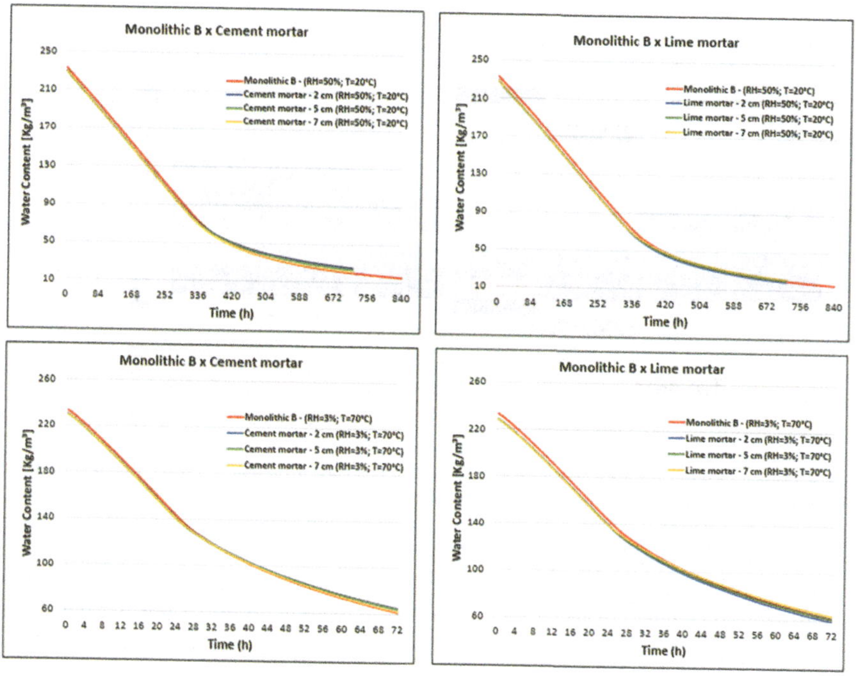

Fig. 6 Water content graph along brick B of simulated transport on the drying process for perfect hydraulic contact (cement × lime)

Figure 6 shows that the sample with cement mortar perfect hydraulic contact with interface positioned at 2 cm had a lower drying rate and the sample with interface positioned at 7 cm had a higher drying rate. The opposite was observed for samples with lime mortar perfect hydraulic contact (see Fig. 7).

In resume:

- The water transport in the absorption process and in the drying process showed similar behavior for the multilayer sample with mortar with higher permeability and lower resistance to water vapor diffusion than the brick. However, the water content is transported at a rate approximately ten times slower for the drying process.
- In all samples, the second layer of brick A and brick B exposed to ambient conditions have lower water contents, which characterizes greater drying in the region. These results were proven for the experimental study by Azevedo (2019).
- The brick A samples with perfect hydraulic contact indicate that the greater the interface placement distance from the base, the shorter the time required for the drying process in the samples.
- The brick B samples with perfect hydraulic contact with cement mortar indicate that the further away from the saturated wet front the material interface is positioned, the more capable it will be to effectively transport the liquid water to the

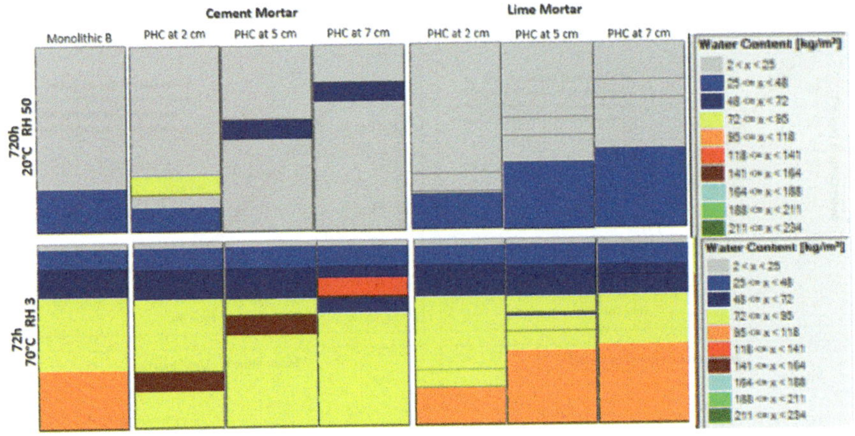

Fig. 7 Water content on the surface of samples with brick B at the end of simulated transport on the drying process for perfect hydraulic contact (cement × lime)

surface where evaporation occurs. This result is due to the greater water vapor diffusion resistance in concrete in Phase II of drying, acting as a water barrier in the sample with interface positioned at a smaller distance from the saturated wet front.

- In the brick B samples with perfect hydraulic contact with lime mortar, the lime layer slightly inhibits the drying process through the brick due to the lower liquid transport coefficient and lower water vapor resistance of the lime compared to brick B.
- The further distant from the saturated wet front the interface is positioned, the longer the drying time in the samples with perfect hydraulic contact with lime mortar.
- The water absorption coefficient and water vapor diffusion resistance of the mortar as well as the permeability of the brick are the main influential properties in an effective drying process.

4.2 Moisture Transport Across Air Space Interface—Drying Process

Figure 8 shows the water transport during the drying process simulation for the brick A samples with air space interface. The water content in the samples is transported uniformly through a linear curve for all samples. The decline in water flow in the samples from a point considered as the interface evidences that an air space interface reduces the transport capacity of water content in the sample, characterizing it as a hydric resistance.

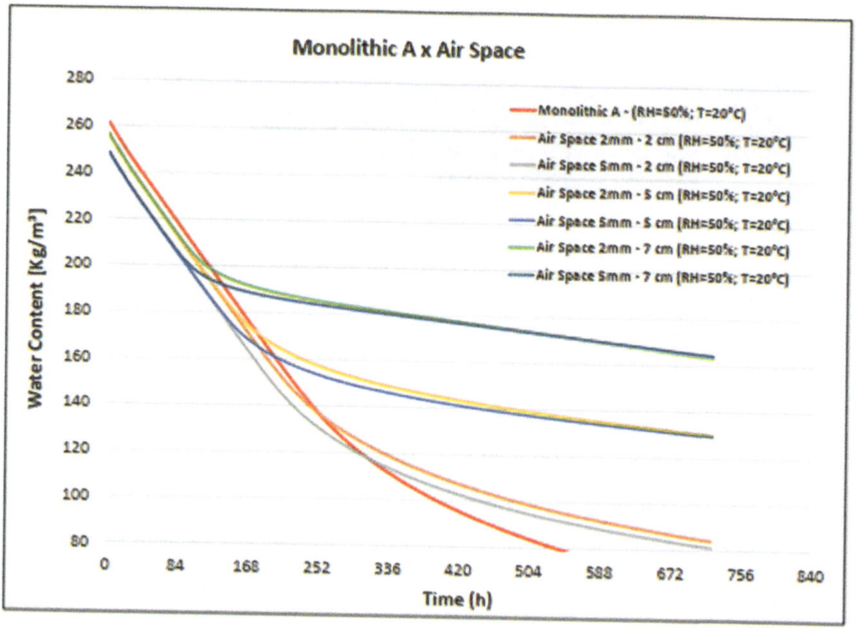

Fig. 8 Water content graph along brick A of simulated transport on the drying process for air space interface

Analyzing Fig. 8 it is possible to observe a long transport during the capillary transport phase for the monolithic sample A. The samples with air space interface located at 7 cm from the base had a faster drying rate reduction than the other samples. It can be seen that the further the distance of the interface from the sealed base (without contact with external environmental conditions and with higher moisture concentration in relation to the top surface), the greater the resistance caused by the interface and the lower the drying rate in the samples.

The reduction in drying rate is greater for samples with a 2 mm air space interface compared to samples with a 5 mm air space interface. However, there is an approximation in the drying rate for the two air interface configurations after a few hours of the drying process.

Figure 9 shows the water content on the surface of brick A samples with 2 mm and 5 mm air space interface at different heights at the end of the 720 h simulated drying process. The samples with a 5 mm air space interface showed lower water content on the surface.

Figure 10 shows the water transport during the drying process simulation for brick B samples with air space interface. The water content in the samples is transported uniformly through a linear curve for all samples. The decline in the water flow in the samples from a point considered as the interface evidences that presence of an air space interface reduces the water content transport capacity in the sample, characterizing it as a hydric resistance.

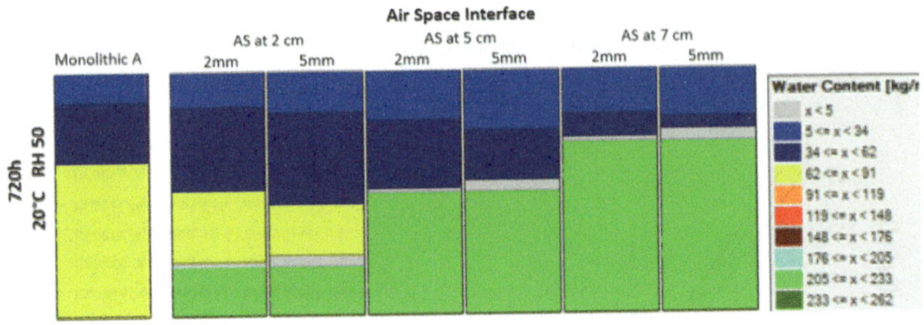

Fig. 9 Water content on the surface of samples with brick A at the end of simulated transport on the drying process for air space interface

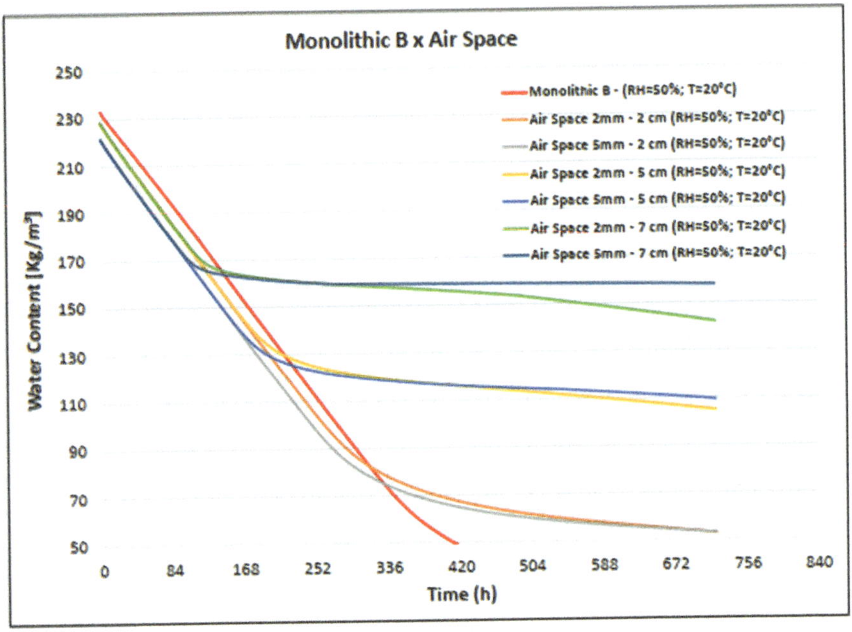

Fig. 10 Water content graph along brick B of simulated transport on the drying process for air space interface

Analyzing the graph (Fig. 10) observed a long transport during the capillary transport phase for the monolithic sample B. The samples with air interface located at a height of 7 cm from the base had a faster drying rate reduction than the other samples. It can be seen that the further the distance of the interface from the sealed base (without contact with the external environmental conditions and with higher moisture concentration in relation to the top surface), the greater the resistance caused by the interface and the lower the drying rate in the samples.

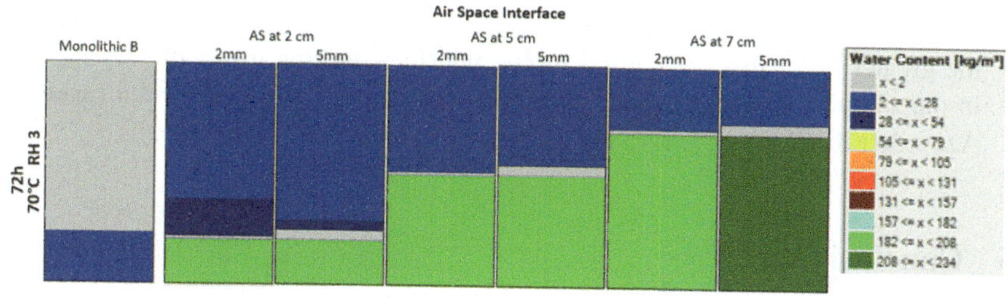

Fig. 11 Water content on the surface of samples with brick B at the end of simulated transport on the drying process for air space interface

The reduction in drying rate is greater for samples with a 2 mm air space interface compared to samples with a 5 mm air space interface. However, there is an approximation in the drying rate for the two air interface configurations after a few hours of the drying process. Yet, by the end of the simulation, the samples with 2 mm air space interface will show an increase in drying rate compared to the 5 mm air interface sample. This explains a higher water vapor permeability caused by the sample assembly, since the 2 mm air layer has a higher water vapor diffusion resistance compared to the 5 mm air layer, but the more permeable brick B layer is thick in the 2 mm air space sample.

Figure 11 shows the water content on the surfaces of brick B samples with 2 mm and 5 mm air space interface at different heights by the end of 720 h drying process simulation. The sample with 5 mm air space interface positioned at 2 cm showed lower water contents on the surface.

In resume it was possible to observe:

- Significant change was observed in the curves for samples with air space interface in comparison with the curve for monolithic samples in the drying process. This result this result was verified in the experimental result of Azevedo (2019).
- The brick B samples with air space interface, as well as the monolithic sample B, showed a faster drying rate in Phase I. The higher permeability of brick B compared to brick A and mortars explains this fact.
- The water transport behavior in the drying process with air space interface in brick A samples were similar to the water transport behavior in the drying process for samples with brick B.
- The 2 mm air layers caused a higher water resistance in water transport through the air space interface.
- The brick permeability will influence the resistance caused by the air space interface in water transport in the drying process. The more permeable the second brick layer is, the smaller the effect caused by the water vapor diffusion resistance of the air layer.

- It can be seen that the further the distance of the interface from the sealed base, the greater the resistance caused by the interface and the lower the drying rate in the samples. This result this result was verified in the experimental result of Azevedo (2019).

5 Conclusions

The development of this study made it possible to verify with the numerical calculation program WUFI-2D, the moisture transport behavior through samples with perfect hydraulic contact and through samples with air space interface. In this work, an attempt was made to numerically simulate the samples studied in Azevedo's (Maroulis and Saravacos 2003) experimental research with the aim of validating the results found in the laboratory and resorting to further analysis with the results obtained in the simulation through the resources made available by the software application.

The most significant results from the simulations and which contributed to the experimental study validation are the following:

- The surface exposed to environmental conditions for the water transport direction in the drying process will show lower water content.
- The further distant from the saturated wet front the interface is positioned, the longer the drying time in the samples with perfect hydraulic contact with lime mortar. This result this result was verified in the experimental result of Maroulis and Saravacos (2003).
- The water absorption coefficient and water vapor diffusion resistance of the mortar as well as the permeability of the brick are the main influential properties in an effective drying process.
- Significant change was observed in the curves for samples with air space interface in comparison with the curve for monolithic samples in the drying process. This result this result was verified in the experimental result of Maroulis and Saravacos (2003).
- The thinner air layer causes a higher water resistance in water transport through the air space interface.
- The brick permeability will influence the resistance caused by the air space interface in water transport in the drying process. The more permeable the second brick layer is, the smaller the effect caused by the water vapor diffusion resistance of the air layer.
- Although the software physics neglects the discontinuity effect caused by the interface between the material layers, the program proved to be helpful for studies of effective assemblies and solutions for moisture transport.

References

Azenha M (2009) Numerical simulation of the structural behavior of concrete since its early ages, Ph.D. dissertation, Department Civil Engineering, University of Porto

Azevedo AC (2019) Interface influence on moisture transport in building components, PhD thesis, Faculdade de Engenharia da Universidade do Porto, Portugal

Bednar T (2002) Approximation of liquid moisture transport coefficient of porous building materials by suction and drying experiments. Demands on determination of drying curve, Proceedings of 6th symposium on building physics in the Nordic countries, Trondheim, Norway, pp 493–500

Crank J (1975) The mathematics of diffusion, 2nd edn. Clarendon, New York

de Freitas VP, Abrantes V, Crausse P (1996) Moisture migration in building walls—analysis of the interface phenomena. Build Environ 31(2):99–108

Delgado JMPQ, Azevedo AC, Guimarães AS (2019) Drying kinetics in building materials and components—the interface influence. SpringerBriefs in Applied Sciences and Technology, Springer

Derdour L, Desmorieux H, Andrieu J (2000) A contribution to the characteristic drying curve concept: application to the drying of plaster. Drying Technol 18(1–2):237–260

Freitas VP, Guimarães AS, Torres MI (2008) Humidade ascensional. FEUP edições, Porto

Hall C, Hoff WD (2009) Water transport in brick, stone and concrete, 2nd edn. CRC Press, London

Iglesias HA, Chirife J (1983) Handbook of food isotherms. Academic Press, New York

Karoglou M, Moropoulou A, Maroulis ZB, Krokida MK (2003) Drying kinetics of some building materials. Drying Technol 23(1–2):305–315

Krischer O, Kroell K (1963) Techniques du séchage, trad. Française de l'éd. 1963 ; Centre technique des industries Aérauliques et thermiques, 91402 Orsay

Kuenzel MH (1995) Simultaneous heat and moisture transport in building components, one- and two-dimensional calculation using simple parameters. Ph.D. thesis, Fraunhofer Institute in Bauphysics, IRB Verlag

Kuitche A, Edoun M, Takamte G (2007) Influence of pre-treatment on drying on the drying kinetic of a Local Okro (*Hibiscus ersculentus*) Variety. World J Dairy Food Sci 2(2):83–88

Mainguy M, Coussy O, Baroghel-Bouny V (2001) Role of air pressure in drying of weakly permeable materials. J Eng Mech 127(6):582–592

Marinos-Kouris D, Maroulis ZB, Kirannoudis CT (1996) Computer simulation of industrial dryers. Drying Technol 14(5):971–1010

Maroulis ZB, Saravacos GD (2003) Food process design. Marcel Dekker, New York, USA

Mendes N, Philippi PC (2003) A method for predicting heat and moisture transfer through multilayer walls based on temperature and moisture content gradients. Int J Heat Mass Transfer 48:37–51

Panchariya PC, Popovic D, Sharma AL (2002) Thin-layer modelling of black tea drying process. J Food Eng 52(4):349–357

Pel L (1993) Moisture transport in porous building materials. PhD thesis, Technical University Eindhoven, The Netherlands

Qin M et al (2009) Coupled heat and moisture transfer in multi-layer building materials. Constr Build Mater 23:967–975

Selih J, Sousa A, Bremner T (1996) Moisture transport in initially fully saturated concrete during drying, transport porous. Media 24:81–106

Van Belleghem M, De Backer L, Janssens A, De Paepe M (2012) Conjugate modelling of convective drying phenomena in porous building materials. Paper presented at the 6th European thermal sciences conference (Eurotherm 2012), 4–7 Sept 2012, UK

Wilson MA, Hoff WD, Hall C (1993) Water movement in porous building materials—XIII absorption in two—layer composite. Build Environ 30:209–219

Non-linear Analysis of Bottle-Shaped Isolated Struts Concrete Deteriorated by Alkali Silica Reactions

F. A. N. Silva, I. S. Lira, A. C. Azevedo, J. M. P. Q. Delgado, A. M. Matos, M. Tahlaiti, and A. Khelidj

Abstract The work discusses the behaviour of isolated concrete bottle-shaped struts affected by internal expansion reactions (ISR). For that purpose, the numerical modelling of damaged concrete was performed using the Concrete Damaged Plasticity Model (CDPM) implemented in ABAQUS and validated the model through Sankovich's tests. A procedure to automatically obtain the concrete plasticity and damage parameters, essential for CDPM, was developed in Matlab. The inputs were the characteristic compressive strength of the concrete, the equivalent length of the finite element mesh and the ratio between the plastic and inelastic compressive strains. The results showed that the CDPM could represent the load-bearing mechanisms of isolated concrete bottle-shaped struts for a range of several stress levels to which these elements may be subjected in the panels investigated. The numerical simulations for different expansion levels consistently captured the expected damage profile of the panels and the load corresponding to the formation of the first crack, the estimated crack opening, and the ultimate load. For the panels investigated, the reduction

F. A. N. Silva (✉) · I. S. Lira
Civil Engineering Department, Universidade Católica de Pernambuco, Recife, Brazil
e-mail: fernando.nogueira@unicap.br

A. C. Azevedo · J. M. P. Q. Delgado · A. M. Matos
CONSTRUCT, Departamento de Engenharia Civil, Universidade Do Porto, Rua Dr. Roberto Frias, S/N, 4200-465 Porto, Portugal
e-mail: antonio.costaazevedo@fe.up.pt

J. M. P. Q. Delgado
e-mail: jdelgado@fe.up.pt

A. M. Matos
e-mail: anamatos@fe.up.pt

M. Tahlaiti
ICAM, School of Engineering Nantes, GeM, CNRS UMR 6183, Research Institute in Civil Engineering and Mechanics, Centrale Nantes, France
e-mail: mahfoud.tahlaiti@icam.fr

A. Khelidj
Laboratory GeM—CNRS UMR 6183, Research Institute in Civil Engineering and Mechanics,, Centrale Nantes, France
e-mail: abdelhafid.khelidj@univ-nantes.fr

J. M. P. Q. Delgado (ed.), *Building Pathologies: Experimental Campaigns and Numerical Procedures*, Building Pathology and Rehabilitation 25,
https://doi.org/10.1007/978-3-031-17061-4_3

observed in the failure load reached values close to 70%, the increase of the tensile plastic deformation was more than 60%, and the maximum crack opening can reach an increase of 113% when compared with those observed experimentally in panels without internal swelling reactions.

Keywords Numerical modeling · Nonlinear analysis · Bottle shaped concrete struts · Internal concrete swelling reactions · Plasticity damage models

1 Introduction

Nearby of sudden changes in geometry or loading, the strain distribution profile doesn't follow a linear evolution law. These regions are represented by hypothetical truss models in which the connecting struts represent the compressed concrete and the ties represent the tensile reinforcements, joined at points called nodes.

Concrete pile caps are volume structures that distribute loads of the columns to deep foundations elements, such as piles and piers. In the design of this structural element, it is common to use the strut and tie method, and, in general, bottle-shaped struts are formed inside this element, as can be seen in Fig. 1a. Bottle-shaped compression struts are wider in the middle than at their ends since the width of the concrete available for stress spreading at this location is bigger. The dotted line in Fig. 1b represents the effective limit of the strut.

Cases of early deterioration of foundation blocks in residential buildings and concrete bridges in the Metropolitan Region of Recife have been reported with relative frequency in the last decade. This process usually starts with the occurrence of a horizontal crack with a wide opening on the lateral faces of the element located approximately 30 cm from the upper face of the block (Silva 2007; Gomes 2008). This occurrence is usually attributed to concrete expansions, resulting from Internal Expansion Reactions (IER), most frequently the Alkali-Silica Reaction (ASR).

Internal expansion reactions are important issues in the early degradation of concrete structures. For the development of these reactions, cement alkali content,

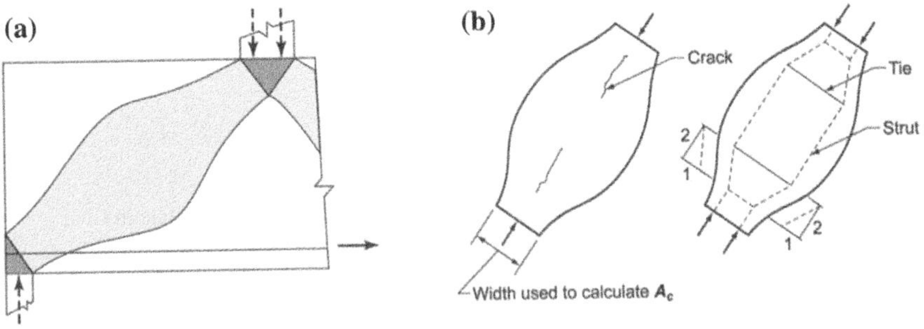

Fig. 1 **a** Bottle-shaped strut and **b** transverse tensile forces

reactivity of coarse aggregates and/or the existence of a source of moisture are necessary and sufficient conditions: Bangert et al. (2004), Lindgård et al. (2012).

The occurrence of ASR in a concrete element can generate an important reduction in its strength and deformation properties, with more pronounced effects in its tensile strength and modulus of elasticity than in its compressive strength (Hobbs 1988; Pleau et al. 1989; Reinhardt and Mielich 2012; Mielich et al. 2015; Sobrinho 2012; Sanchez et al. 2014a, b; Diab et al. 2020; Jiang et al. 2020). In the specific case of the reduction of the tensile strength of the concrete, its consequence in the load capacity of the pile cap blocks is relevant because this reduction can cause a cracking process in the struts that reduces the element load capacity.

1.1 Brief Review on Alkali-Silica Reaction

Alkali-silica reactions (ASR) are responsible for one of the major damage in concrete structures, such as, pile cap blocks, dam bridges, concrete walls, pavements and nuclear power plants. These problems entail the expenditure of a large amount of money to carry out retrofit works, so the scientific community has been showing a growing interest in this subject.

Among the damage caused by ASR we have the expansion and cracking of the structures that lead to a decrease in the durability, strength and modulus of elasticity of these same structures (Rajabipour et al. 2015), as well as affecting the maintenance of concrete infrastructures (Balachandran et al. 2017).

Firstly, it should be noted that alkali-silica reactions require special conditions to develop, namely the existence of a high amount of alkali in the pore solution (pH: 13.0–13.5) and portlandite from cement paste, high moisture content (relative humidity greater than 70%) and some siliceous phases in natural and synthetic aggregates. It is these combined factors that cause the formation of secondary hydration products that can lead to the occurrence of expansions in the concrete (Rajabipour et al. 2015; Balachandran et al. 2017; Bourdot et al. 2016; Dähn et al. 2016).

The concentrations of alkali in cement, derived from clinker raw materials, normally vary between 0.2 and 1.5% of Na_2O equivalent. Thus, depending on the alkaline content, the concrete pore solution can have a pH between 12.5 and 13.5 (Merz and Leemann 2013).

The chemical process of ASR reactions starts when the silica structure is dissolved by the nucleophilic attack of the hydroxide ion (OH^-), which, due to its highly degraded structure, behaves like a hygroscopic silica gel. The resulting product is a gel known as Alkali Silica Gel. This gel, although not harmful to concrete (Ghanem et al. 2010), has a tendency to swell when it absorbs the moisture present in the solution of the concrete pores and, if confined in the matrix, can generate internal tensions. Thus, the greater the amount of moisture absorbed, the greater the internal pressures, which causes the development of microcracks, which in extreme situations, can lead to concrete failure (Rajabipour et al. 2015; Ghanem et al. 2010; Benmore and Monteiro 2010).

Sanchez et al. (2018) showed that the mechanism of internal expansion reactions and the magnitude of microscopic damage to the concrete structure are information that can be obtained through the characteristics of the pattern and extent of the cracking process.

Finally, it should be noted that alkali-silica reactions do not develop quickly, as they take many years to fully propagate, in addition to being destructive only when the alkali concentrations in the pore fluid are very high (MERL Report-09–23 2009). These damages can be observed in concrete structures through the appearance of cracks, misalignment of structural elements, fragmentation and discoloration on the concrete surface, disintegration of the cement paste, surface protrusions and gel exudation (Hobbs 1988; Rajabipour et al. 2015; Balachandran et al. 2017; Molin 1988; Thomas et al. 2011; Swamy 1992).

1.2 Numerical Modelling of ASR

The numerical modelling of the silica alkali reaction in concrete is not a task of simple implementation due to the complexity of the phenomena involved. In general, this modelling is usually conducted from an approach that involves the formation and expansion of kinetic chemical reactions or diffusion or a mechanical fracture approach that comes from the expansion and deterioration of the structure (Murdoch 2015).

Over the past few years, various numerical models of representation of ASR in concrete have been developed (Pignatelli 2012; Bazant and Steffens 2000; Bazant et al. 2000; Ulm et al. 2000; Steffens et al. 2003; Multon and Toutlemonde 2010; Multon et al. 2009; Comi et al. 2009; Grimal et al. 2008a, b; Dunant 2009; Dunant and Scrivener 2010; Alnaggar et al. 2017). Bazant's work (Bazant and Steffens 2000; Bazant et al. 2000) describes the formulation of a model at the mesoscopic level for simulation of the kinetics of chemical reactions and the processes related to diffusion and fracture mechanics of the damage process. Ulm et al. (2000) formulated a chemical–mechanical model in which chemical reactions have a thermo-activated mechanism that accelerates reactions with increasing temperature. In this model, however, the authors did not include the effects of moisture (water) which plays an important role in the kinetics of reactions. In this sense, Steffens et al. (2003) created a model with a similar global approach, but incorporating the effects of dependence on the moisture history of alkali-silica gel. It is a model composed of two stages: (a) the formation of the amorphous gel that is dependent on moisture and (b) an instantaneous process of gel-water interaction, which is influenced by the aging of the gel caused by the action of water. Multon and Toutlemonde (2010) studied the effect of the path travelled by water on the expansion of the ASR reaction when changes in the humidity of the medium are present. The authors concluded that if at any time of the life of a structure affected by ASR, if there is a source of moisture available, the gel already produced can expand and, even in cases where the reaction has already been stopped due to the lack of water, the water source can cause a new formation of ASR gel.

Multon et al. (2009) created a microscopic model to predict the development of ASR expansion in cement mortars with aggregates of different sizes. The model is based on the theory of damage and is able to show the decreasing of mortar stiffness due to the cracking process caused by ASR. The proposed model was able to satisfactorily reproduce the dependence on expansion, on both the size of the aggregate and the alkali content of the medium.

Comi et al. (2009) proposed a thermochemical damage model to simulate the process of expansion and deterioration of concrete strength and stiffness due to alkalis-aggregate reactions. The model assumes that the concrete affected by the reaction behaves like a heterogeneous biphasic material—an expansive gel and a homogenized skeleton of concrete. The microcracking process due to gel expansion is represented by an isotropic damage approach with scalar damage variables, both in tension and compression regimes.

Bangert et al. (2004) created a macroscopic model to describe the process of deterioration of concrete caused by ASR in which the material is conceived as a three-phase mixture: unreacted material, unexpanded material, but already reacted, and expanded material. In this model, the dependence of kinetic reactions and the magnitude of expansion as a function of humidity are taken into account. The authors concluded that the deterioration of concrete imposed by ASR reactions is governed by the simultaneous action of moisture diffusion and kinetic reactions, which led to an important reduction in load capacity and stiffness of the concrete elements investigated.

Grimal et al. (2008a) presented a discussion about the elementary physical principles of an orthotropic viscoelastic-plastic damage model that includes a chemical pressure induced by alkalis-aggregate reactions. The authors also considered the modelling of the effects of moisture on the development of reactions and drying shrinkage. Grimal et al. (2008b), based on the work of Grimal et al. (2008a), calibrated the parameters of the model using analyses of reinforced concrete beams degraded by internal expansive reactions under mechanical loading and non-homogeneous moisture conditions. The model was able to reproduce the markedly anisotropic expansion, the damage observed during the experimental tests and the displacements measured.

Dunant (2009) developed a model to simulate ASR microscopically using the finite element method with a new strategy to calculate the associated damage. The model takes into account the interaction between aggregates and also the effects of transformation of the expansive gel at the individual level of the aggregates. The author reports that it is possible to model the ASR using the finite element method with few parameters to be adjusted. It also reports that the mechanical consequences of the reaction depend on the mechanical properties of ASR gel, which, in turn, are dependent on the availability of Ca^{2+} and K^{+} ions from the porous solution. As a result of this fact, the author states that the mechanical effects of the reaction are also dependent on the type of aggregate and the type of cement used to produce the concrete.

Dunant and Scrivener (2010) proposed a physical model of representation of ASR that takes into account the effect of gel formation on aggregates on the mechanisms of degradation of the reaction, mainly the damage to the aggregates themselves. According to the authors, the model was able to adequately simulate the mechanisms involved in the reaction. Macroscopic free expansion and degradation of the mechanical properties of concrete were reported to be related to the extent of the reaction.

Alnaggar et al. (2017) proposed a model that couples creep, shrinkage and swelling due to ASR using the Lattice Discrete Particle Model (LDPM). A multi-physical formulation was used to calculate the evolution of temperature, moisture, cement hydration and ASR in space and time. The model was calibrated with experiments available in the literature and the results obtained showed that, even during free expansion, an important degree of coupling exists because the ASR induced expansions were relieved by the mesoscale creep.

In this work it was decided to consider the effects of internal expansion reactions in concrete by adopting a strategy that its occurrence in a concrete element leads to a decrease of the its mechanical properties of strength and deformation. This approach is consistent with several studies that observed this fact in numerical modelling and laboratory tests (Sanchez et al. 2014a, b, 2018, 2017).

2 Materials and Methods

The behaviour of isolated bottle-shaped concrete struts deteriorated by internal expansion reactions was numerically modelled considering linear and non-linear regimes. The Concrete Damaged Plasticity (CDP) model of ABAQUS (2010) was used to represent the damage imposed to concrete by the internal swelling reactions (ISRs). CDP model was first calibrated using benchmark tests in isolated concrete specimens performed by Sankovich (2003), and the effect of ISRs was considered according to Sanchez et al. (2014b). A summary of the geometrical configurations of the specimens tested by Sankovich (2003) used to develop the research described in this work is presented in Table 1. The typical geometry of the investigated elements consists of simple concrete panels with dimensions of 914.4 × 914.4 × 152.5 mm that were partially loaded using steel plate with a dimension of 304.8 × 152.5 × 50.8 mm until rupture, as illustrated in Fig. 2. Figure 3 illustrates a typical specimen elevation for use with Table 1.

To obtain the properties necessary for the characterisation of the damage in tension and compression regimes of concrete, MATLAB (2012) scripts were developed. These scripts were based on the work of Alfarah et al. (2017) and allow obtaining the required parameters with few input data, i.e., the characteristic compressive strength of concrete, the equivalent length of the finite element mesh, and the ratio between plastic and inelastic strain in compression. Table 2 summarizes the data from the two panels used in this research.

Table 1 Summary of specimens

Specimen designation	B (mm)	T (mm)	PL top (mm)	PL bottom (mm)	Reinforced
S1-2	914.4	106.1	106.1 × 304.8 × 50.8	106.1 × 304.8 × 50.8	–
S2-3	914.4	152.4	152.4 × 304.8 × 50.8	152.4 × 304.8 × 50.8	Yes
S2-9	914.4	152.4	152.4 × 304.8 × 50.8	152.4 × 304.8 × 50.8	Yes
S3-1	914.4	152.4	152.4 × 304.8 × 50.8	152.4 × 304.8 × 50.8	–
S3-4	914.4	152.4	152.4 × 152.4 × 25.8	152.4 × 152.4 × 25.8	Yes
S3-10	914.4	254.0	101.6 × 304.8 × 50.8	101.6 × 304.8 × 50.8	Yes
S3-11	1524	152.4	152.4 × 406.4 × 50.8	152.4 × 304.8 × 50.8	Yes
S3-12	1524	152.4	152.4 × 304.8 × 76.2	152.4 × 304.8 × 76.2	Yes

Table 3 exhibits an example of input data used in ABAQUS generated by the MATLAB script. However, two issues must be pointed out. Some difficulties were experienced in obtaining the tensile and compressive damage parameters of concrete that allow the proper use of CDP. The experimental data results were scarce. Thus, a routine was developed that determines the referred parameters, namely, the characteristic compressive strength of concrete, the characteristic length of the finite element mesh to be used, and a given ration between plastic and inelastic strain, in compression. The routine followed the work (Alfarah et al. 2017), besides it has the possibility of obtaining the mechanical properties of concrete from ACI-318 (2019) in addition to FIB (2010). This enabled the availability of two hypotheses for obtaining the mechanical properties of concrete, an aspect that increases the probability of adequacy of the numerical model. The second point respects the concrete mechanical properties change as a result of the ISRs, according to Sanchez et al. (2018, 2017). Thus, a second routine automated the calculation of the concrete properties for each level of internal expansion and generated the inputs for the first routine to perform the operations to calculate the tensile and compressive damage parameters for use in the CDP. This group of routines was converted into a single routine in C++ implemented in ABAQUS to run the models studied.

Since the properties of the CDPM model are dependent on these properties, affecting both inelastic strains and damage parameters in tension and compression regimes, a MATLAB script was developed to automate the calculation of CDPM input parameters. Table 4 Sanchez et al. (2018, 2017) summarizes the data extracted from Sanchez's studies that were used in the numerical simulations of this research, which highlight the results of the ASR (Sanchez et al. 2017).

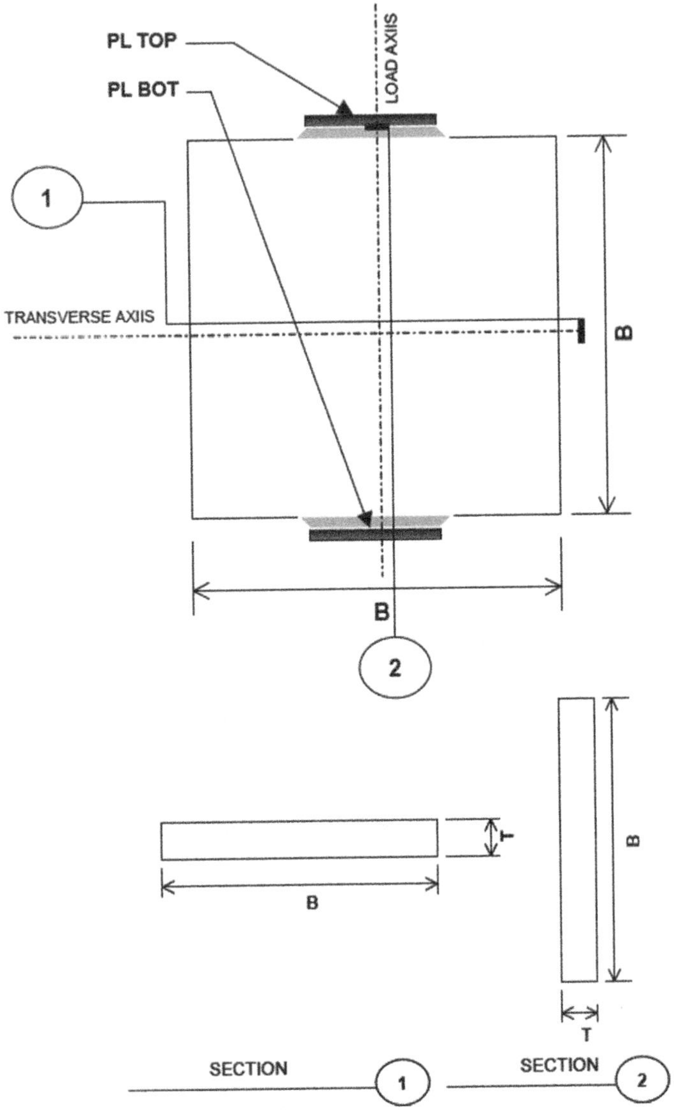

Fig. 2 Typical specimen elevation for use with Table 1

With the CDPM model properly calibrated and validated and with the data of modification of the mechanical properties of the concrete from the expansion level, the following scenarios were analysed: (a) panel fully deteriorated by internal expansions and (b) panel with internal expansions located exclusively on the strut, defined from a linear elastic analysis of the panel (see Fig. 4). The decision to create the

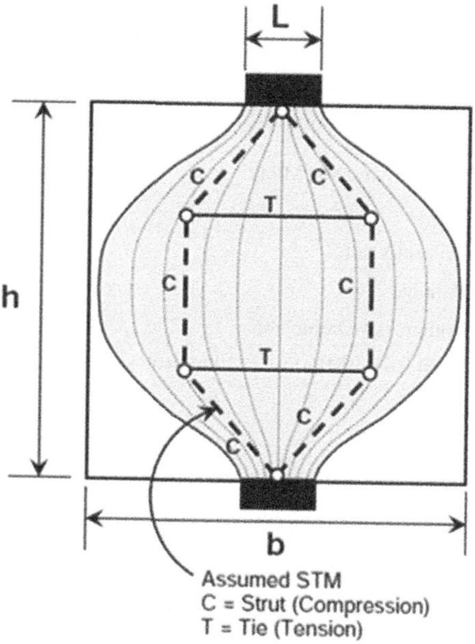

Fig. 3 Bottle-shaped struts and associated strut and tie model

Table 2 Panel results (S1-2 and S3-1)

Panel	Panel dimensions, (mm)	Load plate dimensions, (mm)	f_c, (MPa)	1st cracking load, (kN)	Failure load, (kN)	Principal tensile strain, (%)
S1-2	914.4 × 914.4 × 101.6	304.8 × 50.8 × 101.6	26.41	524.0	706.8	0.0022
S3-1	914.4 × 914.4 × 152.4	304.8 × 50.8 × 152.4	28.96	608.5	873.2	0.0014

scenario (b), despite representing a rare event in usual situations, aims to offer information about the consequences of such an event on the load capacity of the elements studied. This strategy proved justified and relevant because the results obtained are instigating, as will be presented and discussed in subsequent sections.

2.1 Computational Model

The finite element used to model of all panels studied was the C3D8R. The steel plates were modelled with the Discrete Rigid Element—R3D4 family. A structured finite element mesh composed exclusively of cubic elements of edges equal to 50.8 mm was

Table 3 Example of input concrete properties data (S3-1) used in ABAQUS

Elasticity		Plasticity	
Compressive strength, f'c (MPa)	28.96	Dilation angle	5
		Eccentricity	0.1
Elasticity model (MPa)	29,446.22	fb0/fc0	1.16
		K	2/3
Poisson's ratio	0.2	Viscosity	0.0046

Damage			
Compressive behaviour		Compressive damage	
Yield stress (MPa)	Inelastic strain	Damage parameter	Inelastic strain
36.96	0	0	0
21.58	0.00786	0.54	0.00786
14.19	0.0113	0.72	0.0113
8.02	0.0163	0.87	0.0163
1.79	0.0357	0.99	0.0357
Tensile behaviour		Tensile damage	
Yield stress (MPa)	Cracking strain	Yield stress (MPa)	Cracking strain
2.84	0	0	0
2.14	0.0002	0.20	0.0002
1.87	0.0003	0.3	0.0003
1.43	0.0005	0.46	0.0005
1.13	0.0007	0.59	0.0007
0.71	0.0012	0.81	0.0012
0.04	0.0046	0.999	0.0046

used. This decision was taken after the study of several other modelling possibilities and was the one that presented a better computational effort/work ratio of analysis and interpretation of results.

The resulting finite element mesh for panel S1-3 is shown in Fig. 5. The mesh of the plates is shown in Fig. 6. For the modelling of this panel with the upper and lower steel plates, 1008 elements and 1500 nodes were used, totalling 4500 degrees of freedom. It was also considered a contact interaction between the steel plates and the panel surface with a coefficient of friction of 0.3.

The applied load was an uniformly distributed compression stress on the contact surface of steel plate/concrete panel, corresponding to the value of the rupture load of the experimental tests (Sanchez et al. 2014a). The boundary conditions of the top plate were set so that only vertical displacements were enabled. This boundary condition allows the rigid body displacement of the plate in the direction of loading. The same procedures used to impose the boundary conditions on the upper plate were used on the lower plate, but here all three-translation degree of freedom were restrained.

Table 4 Reduction rates for the mechanical properties in accordance with the level of expansion

Distress mechanics	Reference expansion level (%)	Damage results		
		Stiffness loss (%)	Compressive strength loss (%)	Tensile strength loss (%)
ASR 01 (Bangert et al. 2004)	0.04 ± 1	30	60	10
	0.11 ± 1	45	65	15
	0.20 ± 1	60	70	25
	0.30 ± 1	65	80	35
ASR 02 (Lindgård et al. 2012)	0.04 ± 1	37	60	15
	0.11 ± 1	50	65	20
	0.20 ± 1	60	80	25
	0.30 ± 1	67	80	25
ASR 03 (Bangert et al. 2004)	0.05	18	25	2
	0.10	29	40	12
	0.15	38	48	15

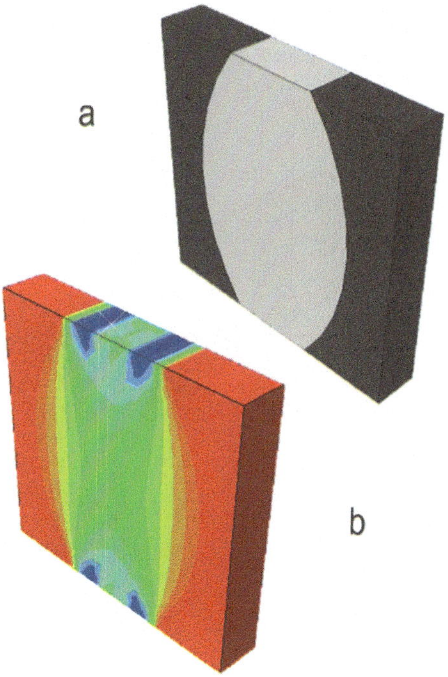

Fig. 4 S3-1 with the strut obtained from linear elastic analysis of the panel highlighted: **a** mesh and **b** numerical model

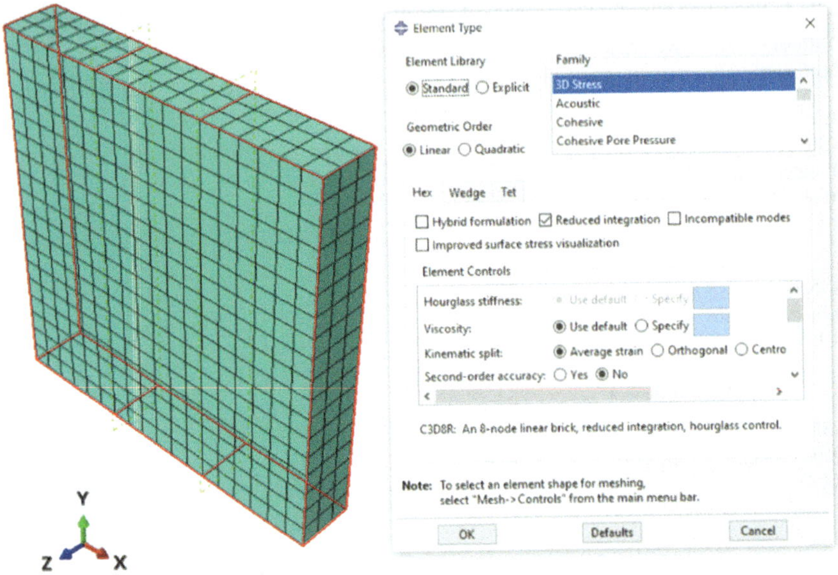

Fig. 5 Finite element mesh (FEM) of S3-1 panel

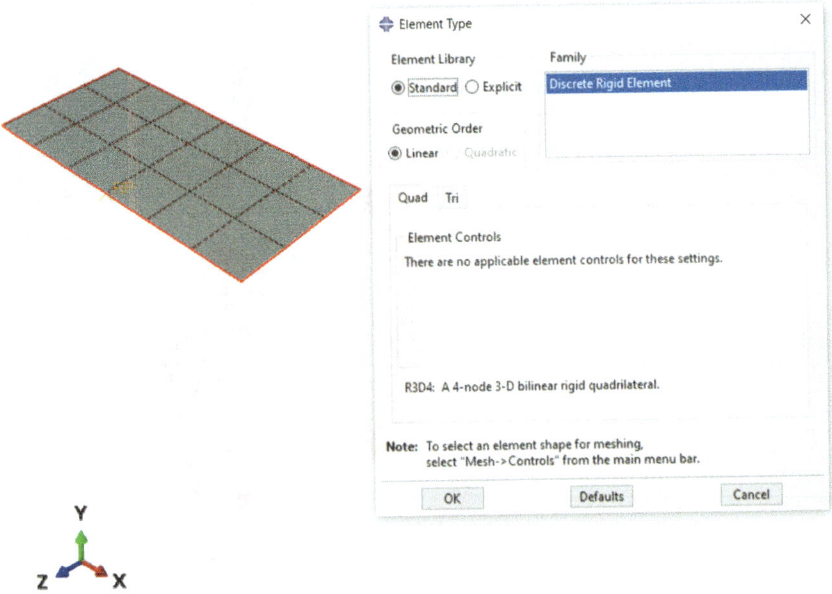

Fig. 6 Finite element mesh (FEM) of the loading plates

3 Results and Discussion

3.1 Validation

Figure 7 shows, for the numerical model and for the laboratory experimental campaign, the tensile plastic deformations of panel S1-2. The numerical and experimental results of the plastic tensile strain profile show a good agreement. This agreement is an excellent indicator of validation of the numerical model developed, as it manages to describe quite accurately the mechanisms involved in the rupture of the panel under analysis, namely the following phenomena: (a) formation of the compression cone at the top and bottom of the panel, and (b) changing the orientation of the plastic deformations in the vicinity of the compression cone.

It should be mentioned that the nodal region has a shape similar to that of an isosceles triangle, with the base twice its height. The described fact was observed both in the experimental campaign and in the numerical results obtained through the developed model. It should also be noted that in this nodal region, the stresses are mainly compressive stresses and that in the vicinity of the region the crushing of the concrete can be observed (see Fig. 7b).

In the numerical model, the tensile damage profile is shown in Fig. 8 next to the failure moment image of panel S1-2. Once again, it is possible to observe the excellent agreement between the experimental and numerical results. The numerical simulation accurately describes the damage that occurred on the side faces of the panel at the break. It should be noted that Sankovich (2003) describes the occurrence of other damage in regions adjacent to the compression edges at the top and bottom of the panel. This phenomenon was also observed through the numerical model developed where a damage of approximately 40% is observed.

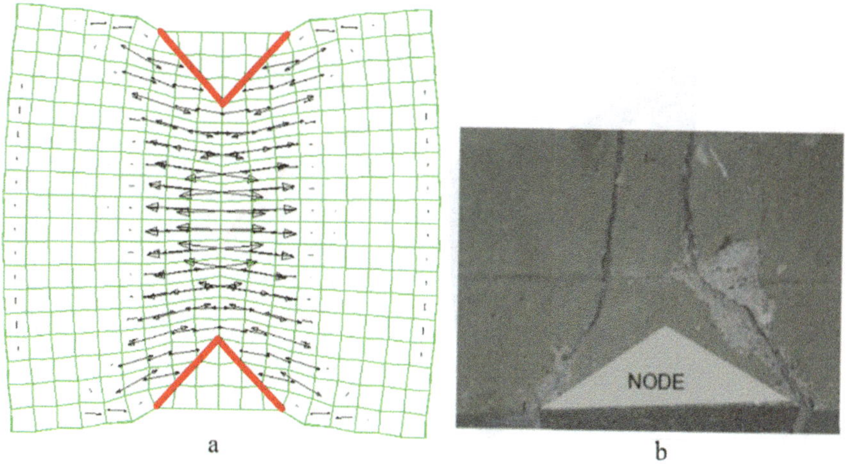

Fig. 7 Experimental S1-2 **a** numerical and **b** plastic tensile strains

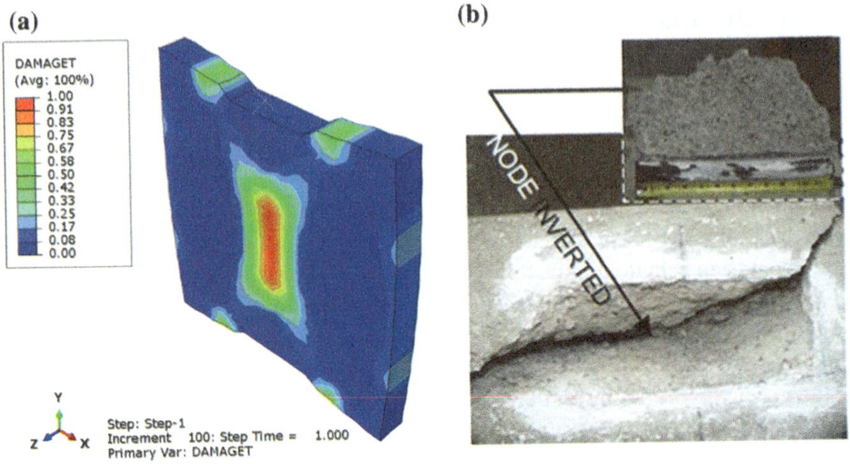

Fig. 8 Tensile damage of the S1-2 **a** numerical model and **b** experimental

The numerical results showed that the tensile damage is more pronounced than the compression damage. This result was validated by the experimental campaign. It is possible to observe that the numerical results present damage in compression of approximately 16%.

Should be mentioned that the profile of compression damage is an important parameter, in numerical models, to validate the location of the compression crushing of concrete samples. As shown in Fig. 9, which presents the numerical values of the profile of compression damage, the damage is concentrated in the vicinity of the nodal region. This observation is in accordance with the experimental results, which presented the same crush location.

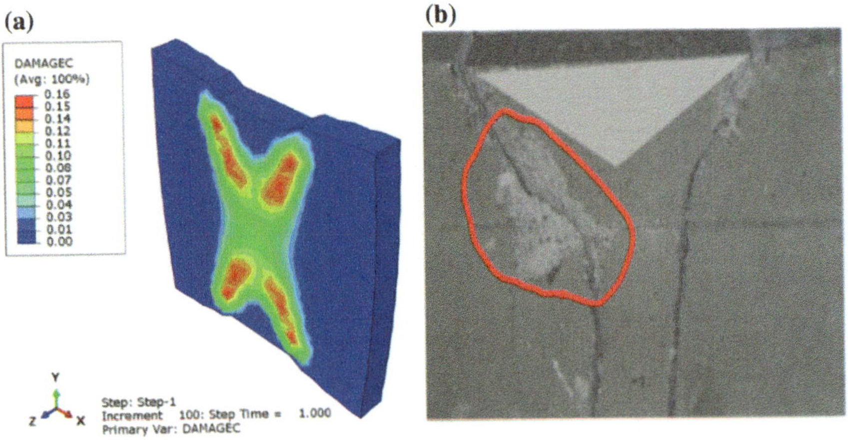

Fig. 9 Compression damage in panel S1-2 **a** numerical model and **b** experimental

Also, the numerical results reproduce with good accuracy the occurrence of vertical tensile stresses that are responsible for the narrowing of the strut observed in the experimental campaign. Figure 10 shows the evolution of STM for different loading levels. It is possible to observe a narrowing of the connecting rod with the increase of the vertical load. This phenomenon is represented in the numerical model by the tensile and compression fields in panel S1-2.

Table 5 presents the experimental and numerical results, namely the first crack loads and the maximum horizontal tensile strains, obtained for the panels analysed in this work. The results showed that the numerical results predicted with good assurance the first crack load of the studied panels, taking into account the complexity of the analysed phenomenon. It is possible to observe that the difference obtained

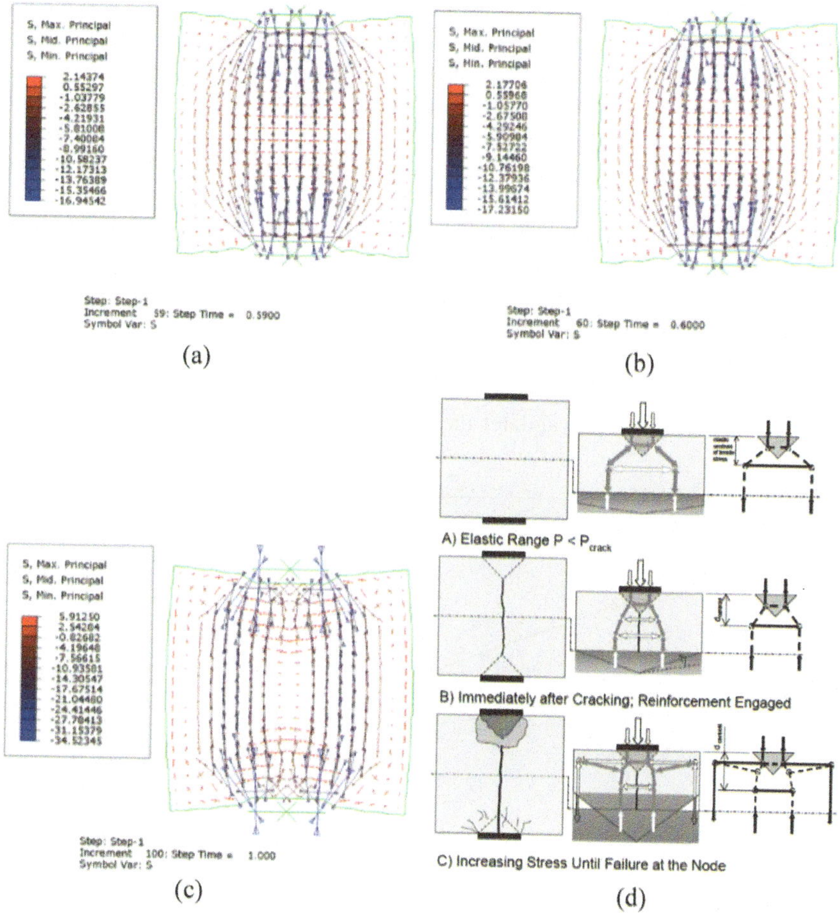

Fig. 10 Numerical model and experimental method used for the evolution of STM for different loading levels **a** elastic range $P < P_{crack}$, **b** immediately after cracking, **c** increasing stress until failure node and **d** experimental method

Table 5 Comparison of numerical and experimental results

Panel	First cracking load (kN)		Error (%)	Maxmum tensile strain (‰)		Error (%)
	Num	Exp		Num	Exp	
S1-2	424.1	524	– 19.06	0.0019	0.0022	– 13.64
S3-1	681.1	608.5	11.93	0.0012	0.0014	– 14.3

between the experimental campaign and the numerical results is 19.06% for the first crack load and 14.31% for the maximum tensile strain.

During the experimental campaign, the authors performed two verifications, justified by the similarity in the instrumentation of different panels, in order to analyse in detail the vertical deformations: firstly, it was compared, for the first crack formation the results of the S3-1 panel, and secondly, another verification was done at failure with data of the S3-10 panel. This comparison is presented in Fig. 11. Finally, it is possible to observe the similarity between the numerical model developed and the experimental method used, for the behaviour of the first crack and failure loads. The results showed that the maximum vertical strains in the centre of the panel decreased to the edges of the panel.

Figure 12 presents the vertical strain results obtained in the experimental models. The measurements were made at half height of the panels along their length. Should be mentioned that the results present in this Figure showed that the responses of the numerical panels were located inside the experimental envelope.

In view of the results presented it is concluded that the numerical models developed were able to efficiently describe the experimental behaviour of the panels investigated, an aspect that validates the numerical models.

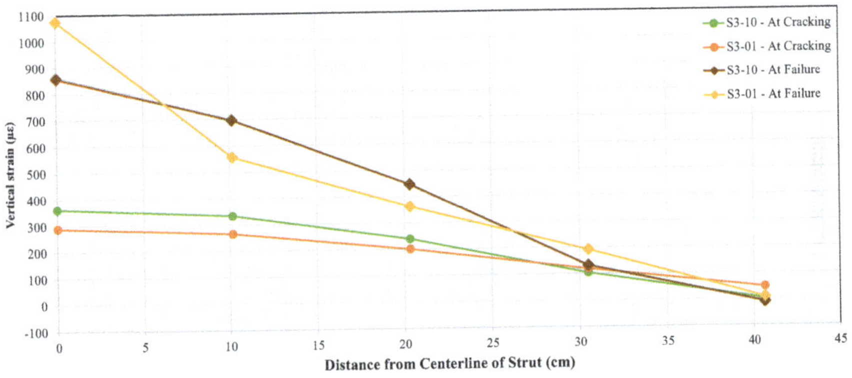

Fig. 11 Comparison between vertical strains at cracking and failure of S3-10 and S3-1 panels

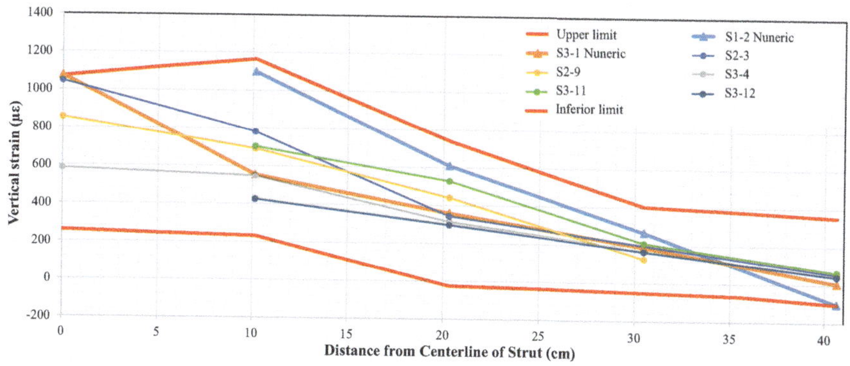

Fig. 12 Comparison between vertical strains at failure

3.2 Panel S3-1 Deteriorated by Alkali-Silica Reaction—ASR

Figure 13 shows the evolution of load and tensile damage with the expansion level at the first crack formation. It can be observed that the panel without internal expansion, hereinafter referred to as intact panel, presented a first crack load of approximately 681 kN and for the expansion level of 0.04% the first crack load was about 297 kN. This means that for a relatively low level of expansion the first crack load decreasing was close to 57%. For the expansion level of 0.30% the first crack load was 175 kN—a reduction of 74.3%. This highlights the important effect of the internal expansion reactions on the overall behaviour of the panel.

When one observes the damage, intact panel exhibited damage of approximately 0.06% at first crack formation and at the expansion level of 0.30% showed a maximum damage at first crack of approximately 0.20%—an increase of approximately 230%. This significant increase shows an important reduction in panel stiffness caused by the internal expansion reaction.

Figure 14 shows the evolution of the load and damage with the expansion level at failure. The failure load of the intact panel was 873 kN and, for the expansion level of 0.04%, its value reached 489.77 kN—a reduction of 44%. The failure load for the expansion level of 0.30% was even lower—281.86 kN—configuring a reduction of approximately 70%. The damage at failure for the intact panel was 74.92% and for the expansion level of 0.30% its value was 96.33%—an increase of 28.58%.

It is also possible to observe in Figs. 13 and 14 that, for relatively discrete expansion levels, an important reduction is already observed, both at first crack and at failure loads. Additionally, it was observed that the effect of discrete expansions was more pronounced at first crack load. For increasing expansion levels, reductions in first crack and failure loads were less pronounced. This fact points to the importance of knowing the levels of expansion existing in concrete parts affected by expansive internal reactions in order to evaluate the need for retrofitting or rehabilitation works.

The failure damage profile for expansion levels of 0.30% is shown in Fig. 15. For an expansion of 0.30%, the fully affected panel exhibited a tensile failure damage

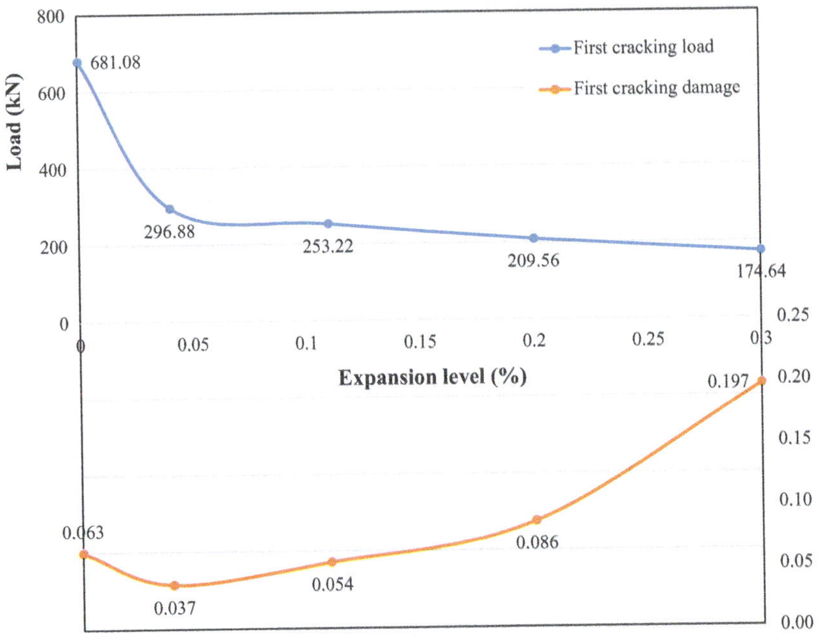

Fig. 13 First crack load and damage for each expansion level

profile at the most distributed in the centre of the panel region. For this panel, damages above 40% were observed throughout the central region. The damage profile on the panel partially affected was more concentrated damage on the peripheries of the strut, adjacent to the compression cone.

The fully affected panels and the panels affected exclusively in the struts presented practically equal load capacity at failure, but with a different distribution of tensile damage profile. The explanation for this fact is that the mechanism of failure of the panels is complex. In fact, the expected failure mode is characterized by the appearance of an approximately vertical tensile crack that gradually increases in opening until a non-ductile failure characterized by crushing the compressive concrete in the vicinity of the nodal region. This last mode of failure was adequately captured by CDPM which showed plastic compression strains in this region, which numerically indicate the beginning of the concrete crushing process.

For an expansion level of 0.30% the profile of plastic tensile strains is different for the three hypotheses studied, as shown in Fig. 16, with the maximum values of 1.28‰, 2.05‰ and 2.06‰, respectively. The differences observed in the profile of plastic tensile strains for the studied expansion levels of the fully affected panel and the affected panel exclusively in the struts points to the importance in choosing the mechanism of allocation of the effects of the expansion in the concrete element

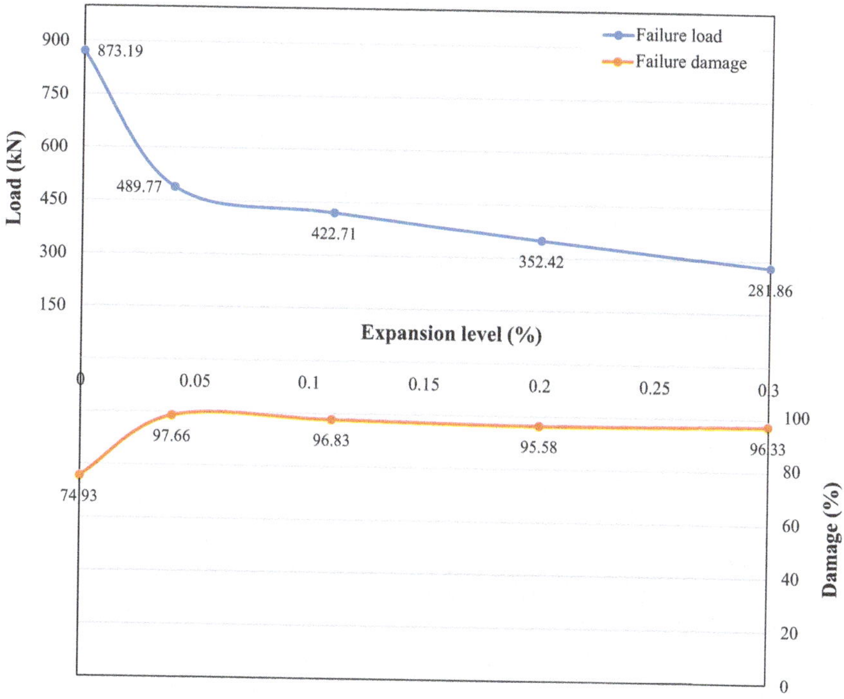

Fig. 14 Failure load and damage for each expansion level

studied. This result shows that the strategy of affecting exclusively the strut is important. For the authors, the choice to affect the strut exclusively is more adequate because of the load support mechanism involved in the concrete element investigated.

Figure 17 shows the evolution of the load and the maximum crack opening at failure along the expansion levels. For panels with an expansion level of 0.04% the maximum crack opening is approximately 1.4 mm and for the panel without expansion the crack opening is approximately 0.77 mm—an increase of 82%. For panels with the maximum expansion level the maximum crack opening is approximately 1.64 mm—an increase of 113% over the maximum panel crack opening without expansion.

Taking into account all the results presented throughout this section, it can be concluded that the effects of reductions in the mechanical properties of concrete are relevant in the loss of load capacity, in the increase of damage in stiffness, in the increase of crack propagation and in the maximum crack opening. The results show that the above-mentioned effects were significant even at small expansion levels. This indicates the importance of preventive measures in the early stages of expansion.

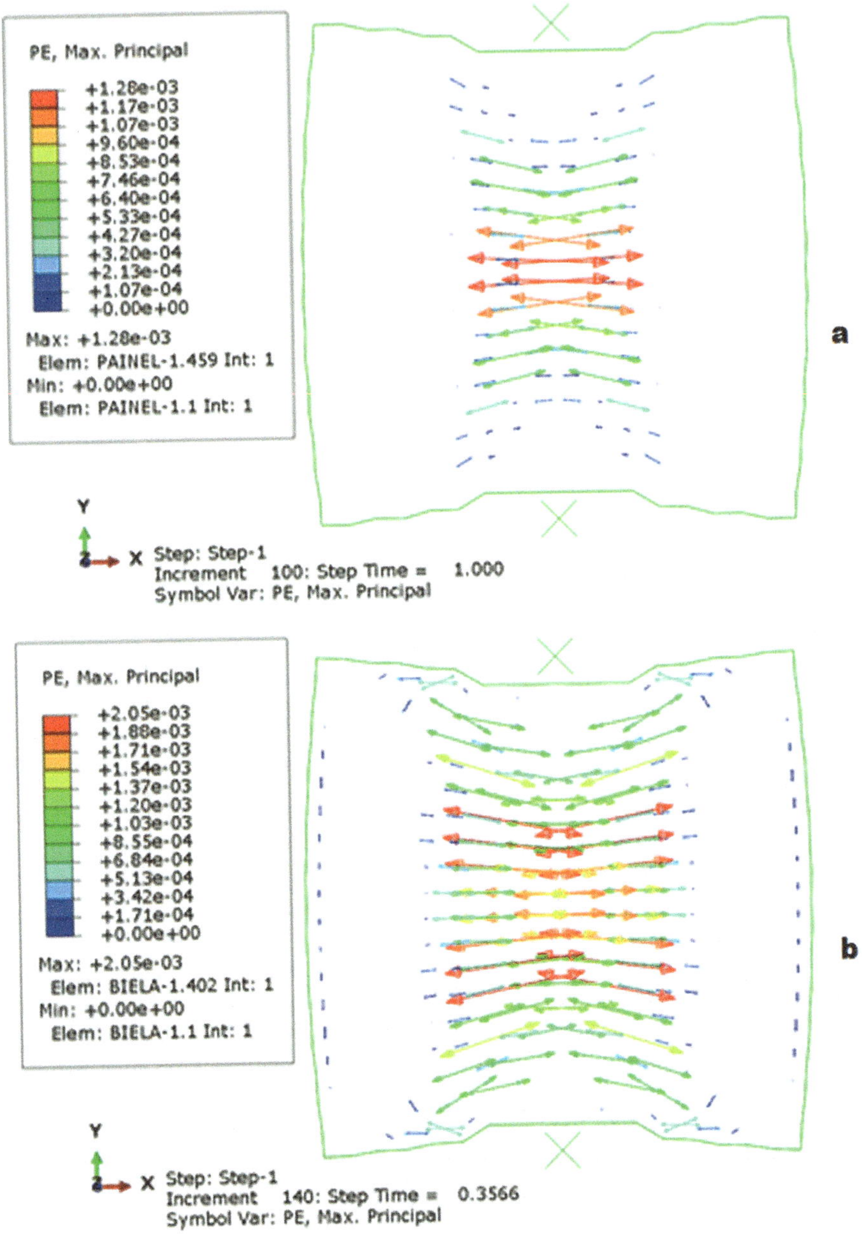

Fig. 15 Panel tensile damage for expansion of 0.30% **a** unaffected, **b** fully affected, **c** partially affected

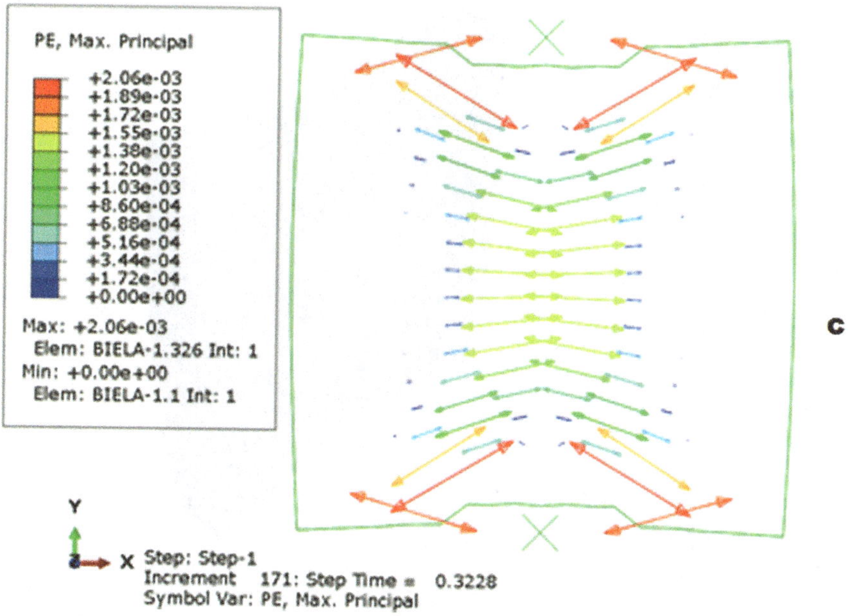

Fig. 15 (continued)

4 Conclusions

The objective of this work was to understand the structural behaviour of isolated bottle-shaped struts deteriorated by internal expansion reactions through numerical modelling with the FEM model. The main conclusions that can be drawn from the research carried out are as follows:

- The ABAQUS CDPM proved to be an efficient numerical strategy for the representation of mechanical phenomena involved in load support mechanisms in partially loaded concrete panels with the formation of isolated bottle-shaped struts. Calibration of the CDPM model with Sankovich's experiments (Sankovich 2003) enhanced the capabilities of the model;
- For the use of CDPM, there is, however, an important difficulty to be overcame—the unavailability of experimental results for a varied concrete property range regarding the plastic strains and damage, both in tension and compression regimes;
- To overcome the difficulties pointed out, a MATLAB script was developed, based on Alfarah's research (Alfarah et al. 2017) that allow obtaining the properties necessary for the use of CDPM in an automated way and with few input parameters—the characteristic compressive strength of the concrete, the equivalent length of the finite element mesh and the ratio between plastic and inelastic strains in compression;

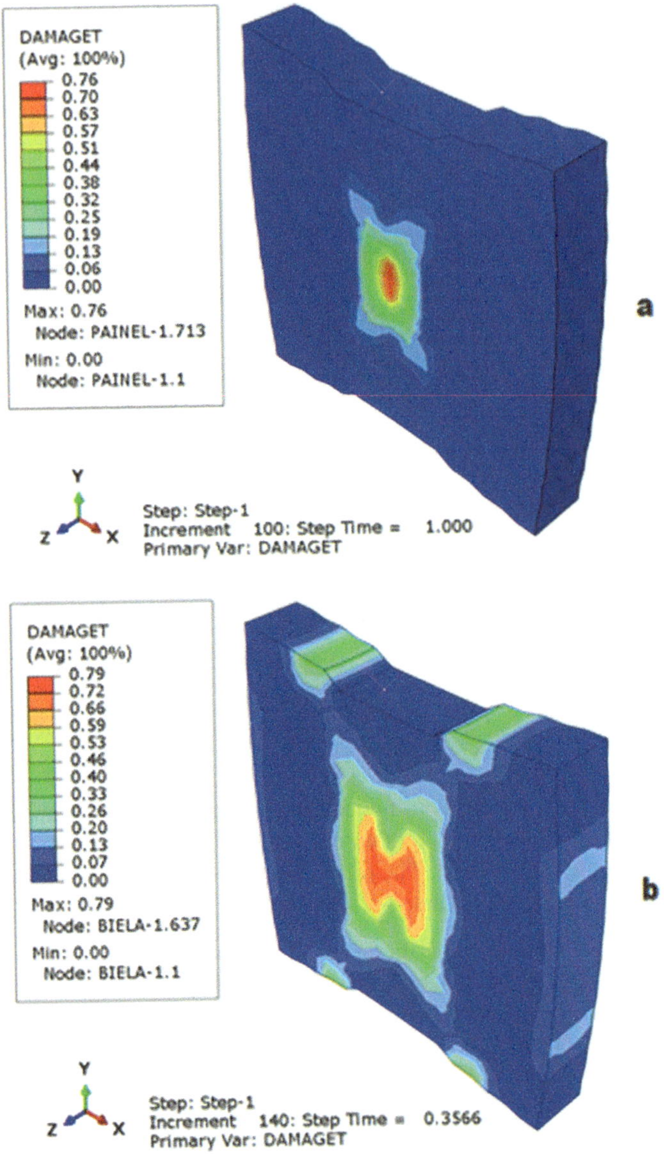

Fig. 16 Plastic tensile strains for expansion of 0.30% **a** unaffected, **b** fully affected, **c** partially affected

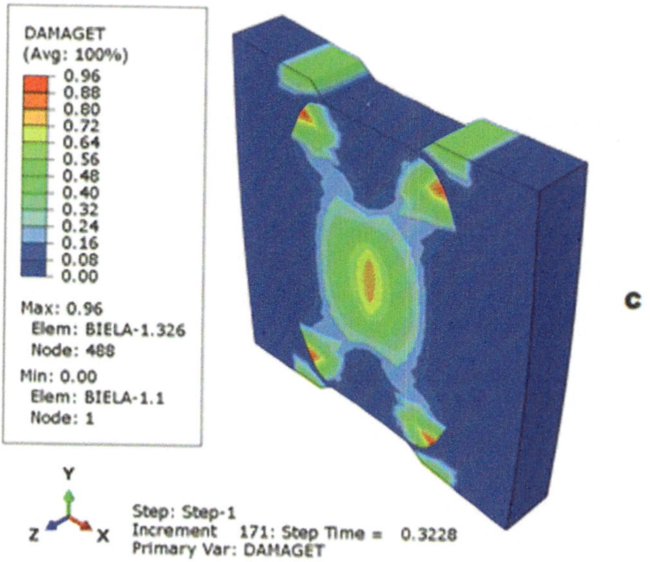

Fig. 16 (continued)

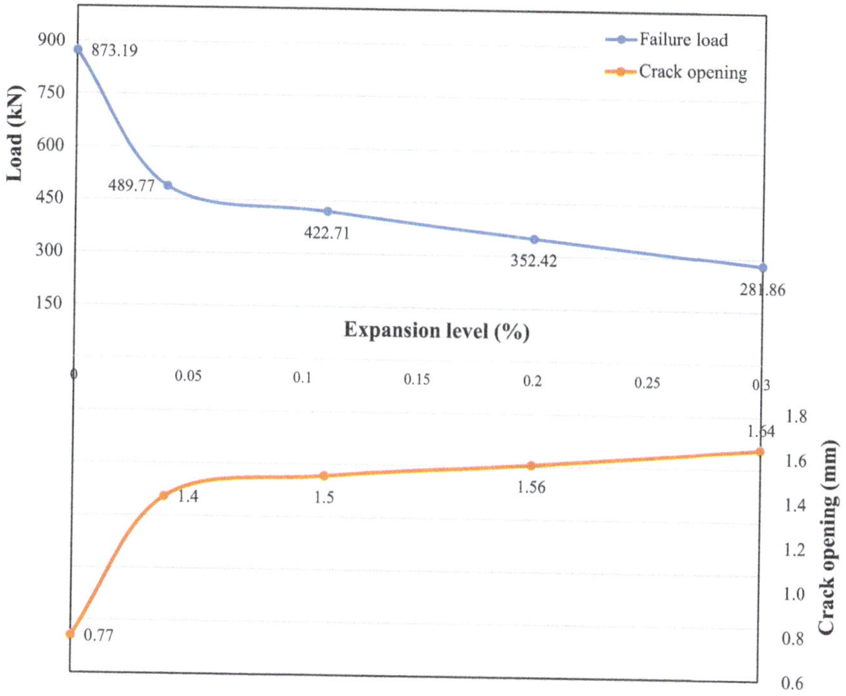

Fig. 17 Load and crack opening at failure for each expansion level

- To perform numerical simulations of concrete with the use of CDPM it is desirable to use a structured and uniform finite element mesh. This is important because the model is relatively sensitive to the equivalent length of the finite element mesh and thus in problems with distinct mesh density will require plastic strains and damages both in tension and in compression regimes, for each region of the model with different equivalent length.

A strategy of numerically simulating the effects of internal expansion reactions on concrete panels partially considering the effect of these expansions on the properties of concrete strength and strain, according to Sanchez et al. (2018, 2017) was investigated in the research and the following results summarize the main findings:

- Slight to moderate internal expansion levels exhibited an important influence on the expected damage profile, the estimation of the first crack load, the estimated crack opening and the load capacity of the investigated panels, both for ASR expansions;
- For the expansions due to ASR, when compared with the values of the intact panel, the following findings stand out:
  - The reduction in the load capacity of the panels studied was 70%;
  - The increase in plastic tensile strain at rupture was 60%;
  - The maximum crack opening increased about 113%.

Considering the early deterioration of foundation blocks in residential buildings in Recife, the current works presented stepwards in modelling and predicting the ASR deleterious expansion effect on Structural Behaviour of Isolated Concrete Bottle-Shaped Struts.

Acknowledgements This work was supported by: Base Funding—UIDB/04708/2020 and Programmatic Funding—UIDP/04708/2020 of the CONSTRUCT—Instituto de I&D em Estruturas e Construções—funded by national funds through the FCT/MCTES (PIDDAC) and by FCT—Fundação para a Ciência e a Tecnologia through the individual Scientific Employment Stimulus 2020.00828.CEECIND and 2021.01765.CEECIND.

References

ABAQUS (2010) ABAQUS user's manual. Dassault Systèmes, Providence, Rohde Island, USA. Simulia Corporation ABAQUS versus 6.10

ACI 318 (2019) Building code requirements for structural concrete and commentary. ACI–American Concrete Institute, USA

Alfarah B et al (2017) New methodology for calculating damage variables evolution in plastic damage model for RC structures. Eng Struct 132:70–86

Alnaggar M, Luzio GD, Cusatis G (2017) Modeling time-dependent behavior of concrete affected by alkali silica reaction in variable environmental conditions. Materials 10(5):471

Balachandran C, Muñoz JF, Arnold T (2017) Characterization of alkali silica reaction gels using Raman spectroscopy. Cem Concr Res 92:66–74

Bangert F, Kuhl D, Meschke G (2004) Chemo-hygro-mechanical modelling and numerical simulation of concrete deterioration caused by alkali-silica reaction. Int J Numer Anal Meth Geomech 28(78):689–714. https://doi.org/10.1002/nag.375

Bazant ZP, Steffens A (2000) Mathematical model for kinetics of alkali–silica reaction in concrete. Cem Concr Res 30:419–428

Bazant ZP et al (2000) Fracture mechanics of ASR in concretes with waste glass particles of different sizes. J Eng Mech ASCE 126(3):226–232

Benmore CJ, Monteiro PJM (2010) The structure of alkali silicate gel by total scattering methods. Cem Concr Res 40(6):892–897

Bourdot A, Thiéry V, Bulteel D, Hammerschlag JG (2016) Effect of burnt oil shale on ASR expansions: a petrographic study of concretes based on reactive aggregates. Constr Build Mater 112:556–569

Comi C, Fedele R, Perego U (2009) A chemo-thermo-damage model for the analysis of concrete dams affected by alkali–silica reaction. Mech Mater 41(3):210–230

Dähn R et al (2016) Application of micro X-ray diffraction to investigate the reaction products formed by the alkali–silica reaction in concrete structures. Cem Concr Res 79:49–56

Diab SH, Soliman AM, Nokken MR (2020) Changes in mechanical properties and durability indices of concrete undergoing ASR expansion. Constr Build Mater 251:118951. https://doi.org/10.1016/j.conbuildmat.2020.118951

Dunant C (2009) Experimental and modelling study of the alkali-silica-reaction in concrete. No. thesis. EPFL

Dunant CF, Scrivener KL (2010) Micro-mechanical modelling of alkali-silica–reaction—induced degradation using the AMIE framework. Cem Concr Res 40(4):517–525

FIB (2010) FIB model code for concrete structures 2010. Ernst & Sohn library, Lausanne, Switzerland

Ghanem H, Zollinger D, Lytton R (2010) Predicting ASR aggregate reactivity in terms of its activation energy. Constr Build Mater 24(7):1101–1108

Gomes EAO (2008) Structural retrofitting works of pile caps foundations deteriorated by alkali aggregate reaction—Recife experience. MSc Thesis, Catholic University of Pernambuco

Grimal E, Sellier A, Le Pape Y, Bourdarot E (2008a) Creep, shrinkage, and anisotropic damage in aar swelling mechanism—part I: a constitutive model. ACI Mater J 105(3):227–235

Grimal E, Sellier A, Le Pape Y, Bourdarot E (2008b) Creep, shrinkage and anisotropic damage in AAR swelling mechanism—part II: identification of model parameters and application. ACI Mater J 105(3):236–242

Hobbs DW (1988) Alkali–silica reaction in concrete. Thomas Telford, London

Jiang Z, He B, Zhu X, Ren Q, Zhang Y (2020) State-of-the-art review on properties evolution and deterioration mechanism of concrete at cryogenic temperature. Constr Build Mater 257:119456. https://doi.org/10.1016/j.conbuildmat.2020.119456

Lindgård J, Andiç-Çakir Ö, Fernandes I, Rønning TF, Thomas MDA (2012) Alkali–silica reactions (ASR): literature review on parameters influencing laboratory performance testing. Cem Concr Res 42(2):223–243

MATLAB (2012) The language of technical computing. The MathWorks, Inc, 2012. Disponível em: http://www.mathworks.com

MERL Report-09-23 (2009) New recommendations for ASR mitigation in reclamation concrete construction. U.S. Bureau of Reclamation, Denver, Colorado

Merz C, Leemann A (2013) Assessment of the residual expansion potential of concrete from structures damaged by AAR. Cem Concr Res 52:182–189

Mielich O, Reinhardt HW, Garrecht H, Giebson C, Seyfarth K, Ludwig HM (2015) Strength and deformation properties of concrete as evaluation criteria for ASR performance tests. Beton- Und Stahlbetonbau 110:554–563

Molin DCCD (1988) Fissures in reinforced concrete structures—analysis of typical manifestations and survey of cases occurred in the state of Rio Grande do Sul, p 238 (in Portuguese)

Multon S, Toutlemonde F (2010) Effect of moisture conditions and transfers on alkali silica reaction damaged structures. Cem Concr Res 40:924–934

Multon S, Sellier A, Martin C (2009) Chemo-mechanical modeling for predition of alkali silica reaction (ASR) expansion. Cem Concr Res 39:490–500

Murdoch R (2015) Creating a simplified model of the alkali-silica reaction in concrete by utilising finite element modelling techniques

Pignatelli R (2012) Modeling of degradation induced by alkali-silica reaction in concrete structures. PhD Thesis. POLIMI, Italy. https://www.politesi.polimi.it/handle/10589/56716

Pleau R, Berube MA, Pigeon M, Fournier B, Raphael S (1989) Mechanical behavior of concrete affected by ASR. In: Elsevier Applied Science (England) (eds) proceedings of the 8th ICAAR, Kyoto, Japan, pp 721–726

Rajabipour F, Giannini E, Dunant C, Ideker JH, Thomas MDA (2015) Alkali–silica reaction: current understanding of the reaction mechanisms and the knowledge gaps. Cem Concr Res 76:130–146

Reinhardt HW, Mielich O (2012) Mechanical properties of concretes with slowly reacting alkali sensitive aggregates. In: Proceedings of the 14th ICAAR conference, Austin, TX, USA

Sanchez LFM, Multon S, Sellier A, Cyr M, Fournier B, Jolin M (2014a) Comparative study of a chemo–mechanical modeling for alkali silica reaction (ASR) with experimental evidences. Constr Build Mater 72:301–315. https://doi.org/10.1016/j.conbuildmat.2014.09.007

Sanchez LFM, Fournier B, Jolin M, Bastien J (2014b) Evaluation of the stiffness damage test (SDT) as a tool for assessing damage in concrete due to ASR: test loading and output responses for concretes incorporating fine or coarse reactive aggregates. Cem Concr Res 56:213–229

Sanchez LFM et al (2017) Overall assessment of alkali-aggregate reaction (AAR) in concretes presenting different strengths and incorporating a wide range of reactive aggregate types and natures. Cem Concr Res 93:17–31

Sanchez LFM et al (2018) Comprehensive damage assessment in concrete affected by different internal swelling reaction (ISR) mechanisms. Cem Concr Res 107:284–303

Sankovich CL (2003) An explanation of the behavior of bottle-shaped struts using stress fields. Master's thesis, University of Texas at Austin

Silva GA (2007) Retrofitting works of pile caps foundations affected by alkali aggregate reaction. MSc Thesis, Catholic University of Pernambuco (in Portuguese)

Sobrinho CWAP (2012) Piles caps of buildings affected by AAR—case study. In: 54° Brazilian concrete congress, Maceió, Brazil (in Portuguese)

Steffens A, Li K, Coussy O (2003) Ageing approach to water effect on alkali-silica reaction. degradation of structures. J Eng Mech 129:50–59

Swamy RN (1992) The alkali-silica reaction in concrete, p 348

Thomas MDA, Fournier B, Folliard KJ, Resendez YA (2011) Alkali-silica reactivity field identification handbook. No. FHWA-HIF-12-022

Ulm FJ, Coussy O, Kefei L, Larive C (2000) Thermo-chemo-mechanics of ASR expansion in concrete structures. J Eng Mech 126:233–242

Strength of Church Towers-Design and Construction

C. Sobrinho, A. Costa, A. C. Azevedo, and J. M. P. Q. Delgado

Abstract The Basilica of Nossa Senhora da Penha is a building of great historical, cultural and religious importance in the city of Recife. The manuscript presents the main results of the reinforcement and structural restoration project of Basilica of Nossa Senhora da Penha, namely, the results associated with the injection techniques of cement grout, the structural safety research results, and the results of the use of carbon fibers. This work presents and discusses the history, current situation, original techniques and strategies used in the development of structural reinforcement design of both towers of the Basilica of Penha Church. Repair techniques poorly designed, conducted in 1981, along with lack of preventive maintenance, leaks and even the growth of bushes embedded in the masonry led to the instability of the towers of the Basilica of Penha Church. This paper, which combining integrated solutions in a historic monument reinforcement project, was initially challenging and became an important case study, possibly one of the first works using carbon fiber reinforcement in masonry. Another important contribution is the insertion of visitable galvanic protection that enables monitoring and possible replacement of sacrificial anodic inserts, keeping the protection active over time.

Keywords Building pathologies · Structural pathologies · Reinforcing masonry · Historic monuments · Reinforcement techniques · Execution strategies · Carbon fibers

---

C. Sobrinho
Housing Technology Laboratory, Pernambuco Institute of Technology, Recife 50740-545, Brazil
e-mail: carlos@itep.br

A. Costa
Pernambuco Polytechnic School, University of Pernambuco, Recife 50120-100, Brazil

A. C. Azevedo · J. M. P. Q. Delgado (✉)
CONSTRUCT-LFC, Departamento de Engenharia Civil, Universidade Do Porto, Rua Dr. Roberto Frias, S/N, 4200-465 Porto, Portugal
e-mail: jdelgado@fe.up.pt

A. C. Azevedo
e-mail: antonio.costaazevedo@fe.up.pt

J. M. P. Q. Delgado (ed.), *Building Pathologies: Experimental Campaigns and Numerical Procedures*, Building Pathology and Rehabilitation 25,
https://doi.org/10.1007/978-3-031-17061-4_4

1 Introduction

The Basilica of Our Lady of the Penha is an important historical monument of the nineteenth century, the only temple in Corinthian style in the form of a Latin cross in Recife, and is located in the federal protection polygon that established the protected national monuments. In 2014, a reinforcement and structural restoration project was elaborated that deserved honorable mention in an international event.

This paper presents the results of studies and injection techniques of cement grout to re-establish the monolithic of very cracked columns, structural safety research results that led to reinforcement with mortar layer replacement and complemented with carbon fibers and studies of cathodic protection of the steel bars in the columns that led to the use of sacrificial inserts (Nascimento 2017; Sobrinho and Costa 2016).

1.1 Characteristics and Situation of the Basilica

The Basilica of Nossa Senhora da Penha is a building of great historical, cultural, and religious importance in the city of Recife. Its architecture has several peculiarities, among which can be mentioned its Latin cross shape, sculptures, and carvings of the styles of the neo-Renaissance, neo-baroque, and eclecticism. For all this importance, this heritage has to be taken care of and studied to remain aesthetically and structurally stable. Initially built in 1656 by the French Calvinist friars, it was paralyzed for over 200 years, undergoing design changes, and was only completed by the Franciscan Fathers from 1870 to 1892 (CECI-Center for Advanced Studies of Integrated Conservation 2021). Figure 1 shows the historical record of the building at the beginning of century XX.

Fig. 1 Historical screen of the basilica in the early twentieth century

Fig. 2 Current view of the tower and the reinforcement in tower columns

Figure 2 shows one of the towers as well as details of the type of reinforcement used in the columns of the towers.

The two towers that house the belfry and the epistle each have eight support columns that support the drum (in masonry) and the arrow (in wood structure) of the roof (in copper plates).

In 1981 reinforcement interventions and filling the windows with cobogos were made in the reinforcements were identified with the inclusion of some columns and bars of steel in elements of reinforced concrete (Araujo 2010).

In 2010, the columns were quite degraded due to corrosion of the central steel bar surrounded by rustic solid brick masonry, assembled, and coated with lime-based mortar. Figure 3 shows some of the degradations observed in the columns.

2 In-Situ Research

Aiming to support the reinforcement project for the two towers, activities were carried out to characterize the compressive behavior in samples taken from the building and a numerical analysis to determine the actions that work in the towers.

a) Armadure degradation   b) Cracked column- next to the auxiliary wood structure   c) Detail of rusted inner steel bar

Fig. 3 Aspects of one of the basilica towers found in 2010

2.1 Physico-Mechanical Characterization

The physical and mechanical characteristics of the building were obtained through inspection by prospecting in areas in the region of the tower of the Epistle, through drill and double disk cutter, with the samples sent to the laboratory of the ITEP-Institute of Technology of Pernambuco.

Attempts to obtain samples using a 4 "diamond drill bit were not efficient, since the need to cut with hydraulic lubrication favoured the solubilisation of the mortar and the brick itself, since both the lime-based mortar and the Not completely burned brick suffered with action of disk movements and water action. In order to obtain samples of the masonry, it was necessary to use a double diamond disc cutter, as shown in Fig. 4.

Samples were cut and rigged in four specimens for achieving the compressive behaviour tests, being used press with displacement control, with a capacity of 300 kN, allowing register the post rupture behaviour.

For the determination of the longitudinal and transverse modulus, deflectometer with the precision of thousandths of a millimetre were installed in the cross-section of load application, and LVDts were used in the measurement of longitudinal and transverse displacements. The composition of Fig. 5 shows the characteristics of the tests on the compressive behaviour of the samples.

Investigations complemented with characterization tests were performed to support structural safety analysis, obtaining the relationship between requesting loads and masonry resistance. Figure 3 shows a record of investigations, samples taken and tests performed.

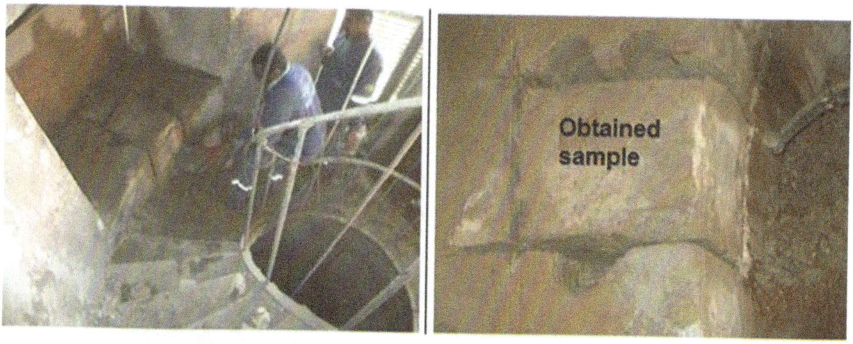

Fig. 4 Characterization of the masonry that makes up the towers of the Basilica: **a** sample removal process using a double diamond disc, and **b** sample removed from masonry sent to the laboratory

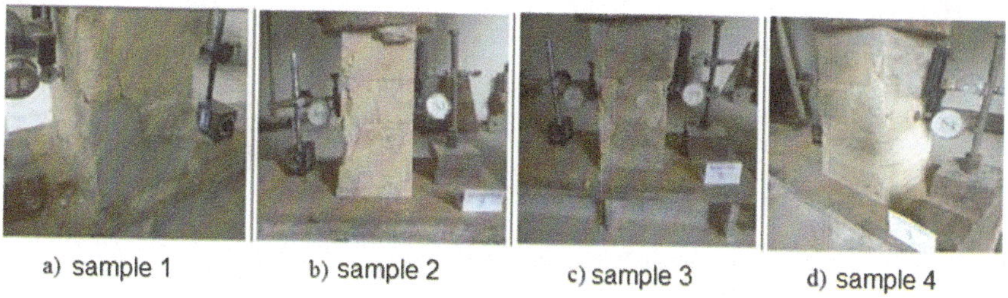

Fig. 5 Compressive behaviour of the assays in the samples

Results of the evaluation of the compressive strength of the samples and applying in the analysis of the determination of the characteristic resistance, according to recommendations of NBR 16868-3 (2020) we obtained $f_{pk} = 1.15$ MPa.

2.2 *Active Stress*

The stressing stresses were obtained based on the numerical modeling based on the finite element method (Mamaghani 2004). The masonry structure was modeled with solid elements of various shapes, the plate elements combined with the membrane elements and the tower cover arrow shell elements. The computer system SAP2000 (NRCS-USDA 2018) was used to obtain the stresses, informing the densities and the characteristics of the compressive behavior (modulus of elasticity and Poisson's ratio).

Figure 6 shows aspects of the solid elements used and the results of the tensions in the various elements that make up one of the towers. This figure shows that the most critical regions of stress concentration are located at the base of the columns,

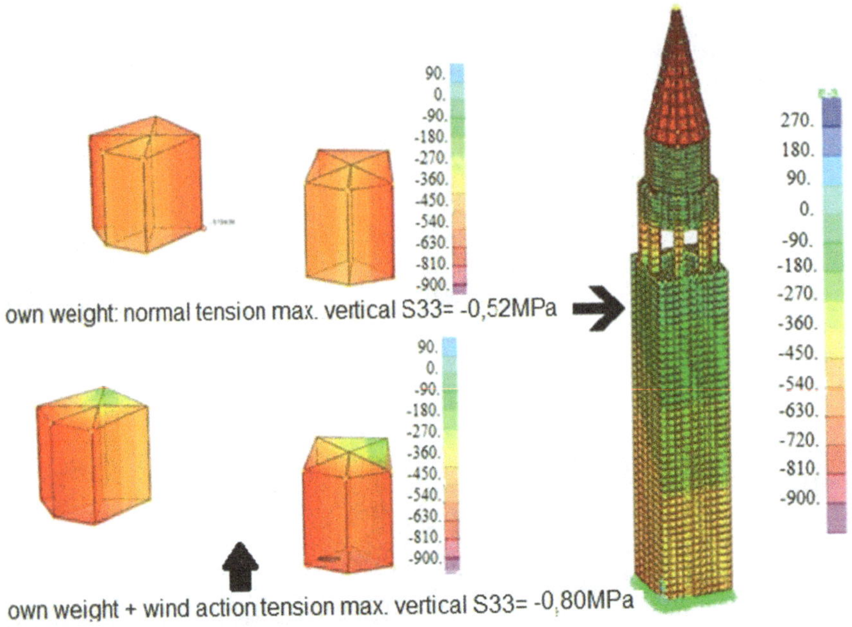

Fig. 6 Results of the numerical analysis of the tower of the basilica

reaching a value of 0.52 MPa due only to its own weight and of 0.80 MPa when considering the combined action of weight and wind action.

These values show that in the actuation of the wind the tensions approached the resistant capacity of the columns, even without considering the safety factors usually existing when sizing and the reduction of the resistance with the cracking of the structure. In this way the temporary reinforcement structures, built with wooden structures in the belfry window beams are acting decisively, avoiding collapse in this region.

The results of these analyzes are very consistent with the situation of the columns of the bell tower and of the Epistle tower, presenting the high state of cracking and indicative of localized ruin.

2.3 *Structural Security Analysis*

Considering the results of the requesting tensions, especially in the region near the bases of the columns, reaching maximum values were between 0.52 MPa and 0.80 MPa and considering that the characteristic resistance of the samples was determined at 0.63 MPa.

These values show that in the performance of the wind the tensions surpass the resistant capacity of the columns, even without considering the safety factors, normally existing when proceeding to a dimensioning. In this way the temporary reinforcement structures, built with wooden structures in the windows of the towers are acting decisively, avoiding collapse in this region.

The results of these analyses are very coherent with the situation that presents the columns of the tower of Basilica, presenting a high state of cracking and indicative of localized ruin. Thus it is concluded that it is extremely necessary to use reinforcement that allows raising at least twice the resistant capacity, thus meeting the normative principles of structural safety.

3 Structural Reinforcement Project

The reinforcement project was developed considering several steps described in detail in the following sub-sections.

3.1 Principles of Reinforcement Design

Composite systems structured with carbon fibers are efficient for the absorption of tensile stresses, avoiding, through the confinement of the section of the axially required pieces, the growth of the transversal deformation of the materials, resulting from the action of the axial load.

The effect of the confinement pressure is to induce a tri-axial stress state in the masonry and under these conditions masonry, or other fragile material, substantially alters its compressive behaviour, both in strength and ductility (Fiorelli 2002).

In addition to the effect of confinement promoted by the use of a carbon fiber and epoxy resin system, the lime-based mortar coating will be replaced with cement-based polymer mortar based coatings and chemical additives (Grande et al. 2011).

3.2 Determination of the Influence of Reinforcement

The compressive characteristics of polymer mortar (Compressive strength of 30.0 MPa and Elasticity Module of 15.0 GPa) in relation to the lime mortar (Compressive strength of 2.0 MPa and Elasticity Module of 0.4 GPa) are substantially larger, being able to overcome the compressive strength by 15 times and the value of the longitudinal modulus of elasticity by more than 35 times (Viapol 2016).

The effect of the confinement, promoted with the use of a system composed of carbon fibers and epoxy resin, can take up to 30% the resistant capacity of a compressed part. Thus combining the effects of coating replacement with confinement on the outer cross sections of the columns provides an increase in the resistive capacity of these elements that make up the towers.

Figure 7 shows the positioning of the columns that present the most critical situations in terms of stress concentration. Practically after the cutting interventions (corners roughing) and rigging, necessary to enable cross-sectional casting of the columns, the estimated areas of masonry and cladding do not change. The evaluation of the active and resistant loads at the base of columns can be estimated in:

(a) Total active load, due to its own weight, considering the numerical modelling: $S_{pp} = 5.20 \times 2630 = 13.676$ ton
(b) Total active load, due to the combined action of own weight and wind: $S_{pp} = 8.00 \times 2630 = 21{,}040$ ton
(c) Resistance of the current masonry, considering the characteristic resistance obtained in the test: $R_a = 6.30 \times 2630 = 16.569$ ton
(d) Estimated strength for reinforced masonry with replacement of lime mortar coating by polymer mortar: $R_{r1} = 6.30 \times 2260 + 300 \times 370 = 127.569$ ton
(e) Estimated strength for reinforcement with the use of the carbon fiber belt: $R_{r2} = 127.569\ 1.20 = 153.082$ ton.

In this way, it can be considered that the proposed reinforcement allows a resistance increase of the columns in 7 times its resistant capacity and if compared to the load acting on the base of the columns due to combined action of own weight and wind action. In this way, the proposed reinforcement presents a security coefficient of the order of 7.0, well above the 2.0 recommended by masonry standards.

4 Reinforcement Procedure

4.1 Restore the Monolithic of the Column Masonry

The masonry of the columns has been severely weakened over time by the oxidation of the center bar, 75 mm (3″) solid cylindrical steel bar located inside the columns and steel plate clamp erroneously placed earlier in an attempt to stop the processes.

In this sense, it was designed the injection cement grout whose measurement was performed using a few additions and additives, with reference to the viscosity measured by the Marsh cone, and low shrinkage (NBR 7681-1 2013). Table 1 shows the dosages surveyed.

The paste (cement grout) represented by sample 7 was taken as a reference to be used on-site, considering the ease of mixing and the results obtained, fluidity of 8 min and 30 s to pour 1.0 L into the cone marsh and low exudation. Figure 8 presents a composition of photos showing the steps of laboratory testing.

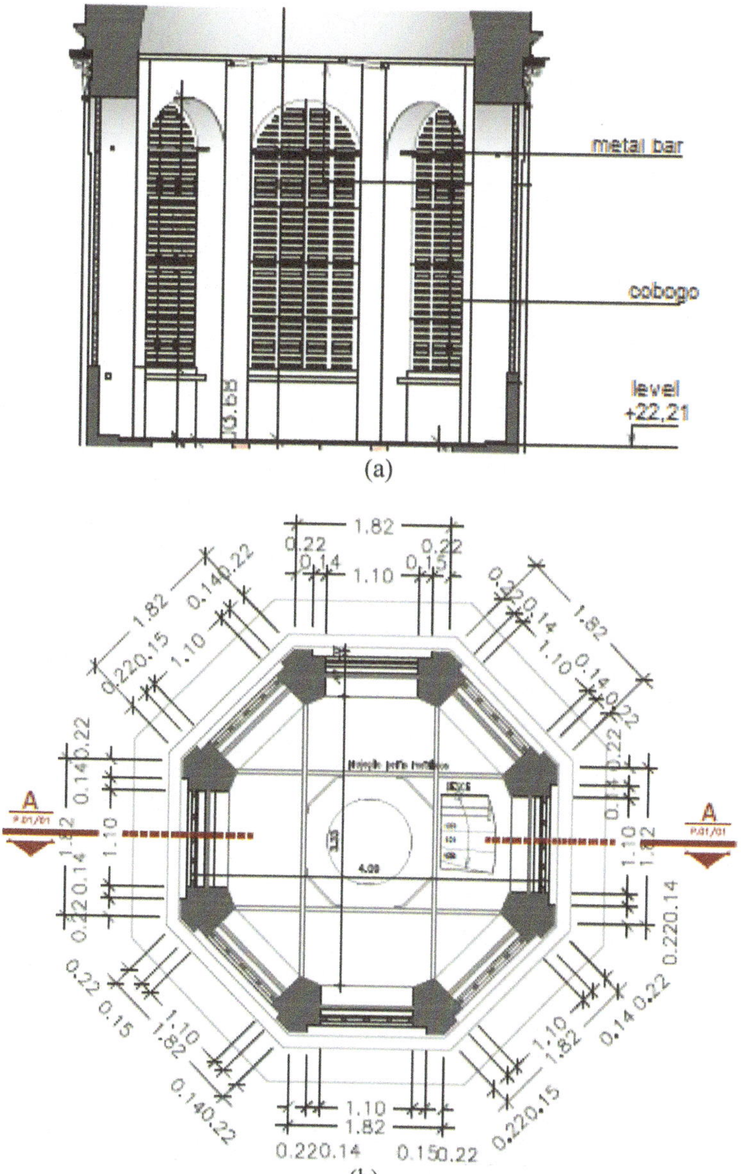

Fig. 7 **a** Drum region and **b** low plant at the base of the columns

Table 1 Results of the evaluated cement grout viscosities

Sample	Cement (g)	Water (g)	Addition (g)		Additive (g)		Time marsh	Comments
1	1000	1000					8:15	Very slump, empty interior
2	1500	750					27:26	Medium slump, dense aspect
3	800	800	400	$CaCO_3$			11:50	Low slump, dense aspect
4	600	600	600	$CaCO_3$			36:37	Very slump, empty interior
5	700	1050	350	Lime-CHII			9:10	Medium slump, dense aspect
6	800	800	200	Lime-CHII			10:88	Very slump, dense aspect
7	1400	700			40	Superplasticizer	8:30	Low slump, dense aspect
8	1400	700	350	Lime-CHII	40		52:10	Low slump, dense aspect
9	1400	700	140	Metakaolin	40		22:10	Low slump, dense aspect

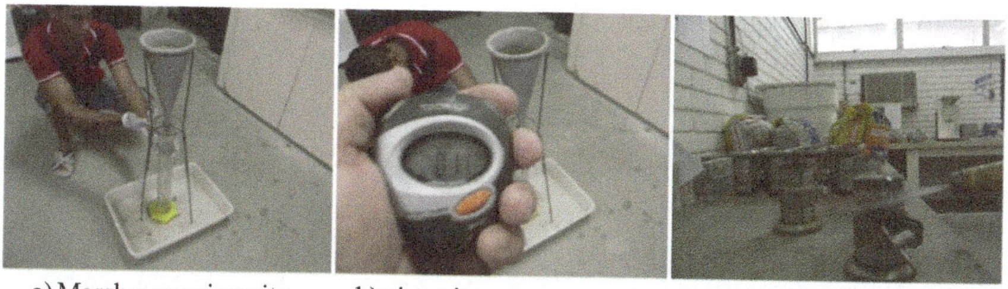

a) Marsh cone viscosity b) viscosity measurement c) fill CPs

Fig. 8 Steps of the injection grout dosing study

a) epoxy seal b) paste mixture c) injection of paste into the trap

Fig. 9 Injection steps in masonry columns

The paste injection procedure in the cracked regions was performed with a hand-operated pump and preparation of the epoxy paste and trap areas. Figure 9 shows the column consolidation step by injection.

4.2 Reinforcement with Carbon Fiber Blankets

Strapping an element under compression allows for increased strength due to the transverse confinement produced when applying loading to the element. Carbon fibers have a high modulus of elasticity (~ 220 GPa) and tensile strength (~ 4 GPa), enabling greater strength gain for the columns (Corum et al. 2000).

The columns, in turn, did not have a cylindrical transverse shape, presenting an angular shape, requiring the preliminary preparation with cutting/thinning and polymeric mortar for subsequent application of carbon fiber blankets. Figure 10 shows steps of applying blanket to columns.

Fig. 10 Carbon fiber application steps: **a** application of saturant on carbon fiber. **b** Saturating being applied and fiber being wrapped in plastic. **c** Fiber application on the column. **d** Fiber fixation on the column

4.3 Cathodic Protection on the Internal Column bar

In order to prevent the oxidation process of the central steel bar inside each column to be advanced, the type of protection that avoided not only stopping for a period, but that could monitor and replace the consumed pellets was studied.

For the design, the electrical resistivity in the columns was measured and the galvanic anodes manufacturer's abacus was used (Chemical 2016).

The tablet was placed in metal boxes fixed to the steel bar by copper wires, after placement the boxes were sealed with silicone, allowing to visualize their behavior over time. Figure 11 shows the steps for measuring the resistivity and placing the prepared pellet boxes on the columns.

a) resistivity measurement b) sacrificial anode (tablet) c) boxes inserted in the column

Fig. 11 Steps sizing and installation of sacrificial anode tablets

5 Conclusions

The situation of the towers of the Basilica of Penha shows signs of concern regarding structural stability. In the evaluation carried out there is no safety reserve, the temporary reinforcement implemented is acting in an effective and full, but new signs of ruin are on display. The proposal of reinforcement presented makes possible not only the removal of the provisional reinforcement the cobogó closing built after the construction of the basilica in the XVIII century. The use of reinforcement on the basis of replacement of lime mortar by polymer mortar and carbon fiber sheeting may be covered by a new layer of mortar based on lime with reconstitution of architectural details and frescoes similar to the original ones.

The reinforcement in carbon fibers and polymer mortar does not suffer degradation with the relative humidity and natural weathering actions, is thus considered durable.

Acknowledgements FUNDARPE- Pernambuco Historical and Artistic Heritage Foundation, PRONEB—Our Lady of Penha Province of Northeast Brazil. This work was supported by: Base Funding—UIDB/04708/2020 and Programmatic Funding—UIDP/04708/2020 of the CONSTRUCT—Instituto de I&D em Estruturas e Construções—funded by national funds through the FCT/MCTES (PIDDAC) and by FCT—Fundação para a Ciência e a Tecnologia through the individual Scientific Employment Stimulus 2020.00828.CEECIND.

References

Araujo RAD (2010) Construction system of the bell towers of the Basilica of Penha. Technical opinion

CECI-Center for Advanced Studies of Integrated Conservation (Accessed in December 2021). http://www.ceci-br.org/obras/penha.htm

Corum JM, Bettiste RL, Lui KC, Ruggles MB (2000) Basic proprieties cross-ply carbon fiber composite. OAK Ridge National Laboratory, Tennessee, USA. www.osti.gov/biblio/777662-basic-properties-reference-crossply-carbon-fiber-composite

Euclid Chemical (2016) Sentinel galvanic anodes. .www.euclidchemical.com/products/construction-products/cathodic-protection/sentinel-galvanic-anode

Fiorelli J (2002) Use of carbon fibres and glass fibres for reinforcement of wooden beams. MSc Thesis, Federal University of Santa Catarina, UFSC, Santa Catarina, Brazil

Grande E, Imbimbo M, Sacco E (2011) Bond behaviour of CFRP laminates on clay bricks: experimental and numerical study. Compos Part B 42:330–340

Mamaghani IHP (2004) Analysis of masonry structures by discrete finite elements method. In: Proceedings of the 4th international seminar on structural analysis of historical constructions, vol 1. Padova, Italy, pp 650–666

Nascimento CRSMS (2017) Procedures for the recovery of the epistle tower of the Basilica of Penha—Recife. Monograph presented specialization course in inspection, reinforcement and maintenance of structures, 74pp, POLI/UPE, Brazil

NBR 16868-3 (2020) Structural masonry Part 3: test methods. Brazilian Association of Technical Standards (ABNT), Rio de Janeiro, Brazil

NBR 7681-1 (2013) Grout for pre-stressing tendons. part 1: requirements. Brazilian Association of Technical Standards (ABNT), Rio de Janeiro, Brazil

NRCS-USDA (2018) Engineering tools and software. www.nrcs.usda.gov/getstarted

Sobrinho CWAP, Costa AC (2016) History, situation and reinforcement of the bell towers of the Basilica of Penha-Recife-Brazil. ALCONPAT J 6(3):200–213

Viapol (2016) Reinforcing manual of carbon fiber reinforced concrete structures. (Website visit 2016). www.viapol.com.br/media/97576/manual-fibra-de-carbono.pdf

Non-destructive and Destructive Tests to Drive Corrective Intervention Procedure of Concrete Elements

A. C. Azevedo, S. Lemos, J. M. P. Q. Delgado, F. A. N. Silva, and C. A. P. Sousa

Abstract This work presents the results of an extensive non-destructive and destructive experimental campaign performed on thirty concrete columns of a 10 multi-storey residential building located in Paraíba, Brazil, built about 30 years ago, to assess the concrete quality to decide the need for retrofitting works. Ultrasound measurements, electrical resistivity, the corrosion potential of the reinforcements, depth of the carbonation front, void index, compressive strength, and capillarity absorption tests were performed. Results indicated the occurrence of corrosion of reinforcement, low concrete compressive strength, unappropriated reinforcement covering, and water absorption level close to 17% void index of about 31%. Tests in drilled concrete cores extracted corroborated the non-destructive test results with average compressive strength significantly low close to 12 MPa, a value significantly lower than the design characteristic strength, i.e., 20 MPa. The joint realization of non-destructive and destructive tests proved to be an efficient strategy for obtaining relevant information for the diagnosis of pathologies in reinforced concrete elements.

Keywords Case study · Experimental campaign · Non-destructive tests · Destructive tests

A. C. Azevedo · J. M. P. Q. Delgado (✉)
CONSTRUCT-LFC, Departamento de Engenharia Civil, Universidade Do Porto, Rua Dr. Roberto Frias, S/N, 4200-465 Porto, Portugal
e-mail: jdelgado@fe.up.pt

A. C. Azevedo
e-mail: antonio.costaazevedo@fe.up.pt

S. Lemos
Reckon Consultoria Em Engenharia, Pernambuco, Recife, Brazil

F. A. N. Silva · C. A. P. Sousa
Civil and Engineering Department, Catholic University of Pernambuco, Pernambuco, Recife, Brazil
e-mail: fernando.nogueira@unicap.br

J. M. P. Q. Delgado (ed.), *Building Pathologies: Experimental Campaigns and Numerical Procedures*, Building Pathology and Rehabilitation 25,
https://doi.org/10.1007/978-3-031-17061-4_5

1 Introduction

Reinforcement corrosion is the main cause and origin of the deterioration of reinforced concrete structures. The reinforcement corrosion mechanism includes that the useful life of a reinforced concrete structure is subdivided into two phases, the initiation period and the propagation period. The initiation period corresponds to the time interval required for aggressive agents, especially carbon dioxide and/or chlorides (CO_2, Cl^-) to penetrate the concrete covering (through the solution of the pores of the cement matrix) to the surface of the reinforcement and the conditions for the destruction of the passivation film are created (rupture of the protective layer of Fe_2O_3). The propagation period corresponds to the time of evolution of the corrosive process, in which the electrochemical reactions that occur in the pore solution favor the dissolution of iron (oxidation), originating corrosion products, until the limit in which the structure reaches a degree of degradation accentuated, making its repair essential or, in later cases, its demolition.

It is well known by the scientific community and concrete technologists that the presence of micro-cracks in concrete structures allows aggressive agents to enter their interior. Once inside the structure, these aggressive agents initiate the corrosion process on the reinforcements in conjunction with water (electrolyte) and oxygen, a phenomenon similar to the micro and macro corrosion of a cell.

Among the factors involved in the corrosion process are the environment to which the structure is subject, the conditions of the structure, its efficient or poorer coverage, the speed of penetration of aggressive agents into the concrete and even the materials used in the concrete.

Among the most aggressive agents that cause the deterioration of reinforced concrete structures, causing a significant portion of the performance loss, we have the transport of chloride ions that cause the destruction of the protective passive film formed on the surface of the embedded steel bar due to the alkalinity of the middle (Metha and Monteiro 2014).

In recent years, the literature shows that several methods have been developed for accelerated simulation of the chloride transport process in concrete (Baroghel-Bouny et al. 2007; Setzer et al. 2004; He et al. 2012; Zhang and Gjørv 1994; Tang 1996), and the drying-wetting cycle test is a good and efficient strategy to evaluate this occurrence (Dantas Neto 2004; El-Din et al. 2017; Dinakar et al. 2013; Ray et al. 2012).

Qualitative techniques, such as the corrosion potential method, which evaluate the thermodynamics of the corrosion process, however do not provide data on the phenomenon associated with the kinetics of reinforcement, continue to be one of the most common electrochemical tools to assist in inspection, monitoring and diagnosis of corrosion of reinforcement of concrete structures (Setzer et al. 2004; Andrade and Alonso 2001; Assouli et al. 2008; Medeiros-Junior and Lima 2016; Liam et al. 1992; Helene 1993a; Broomfield et al. 2002; Elsener et al. 2003; Feliu et al. 2005; Helene et al. 2006; Poupard et al. 2006; Castro-Borges and Ordaz 2009; Medeiros et al. 2012, 2013a, b; Castro-Borges et al. 2012). Furthermore, the corrosion potential method has been commonly used to evaluate the efficiency of repair systems, measure the re-passivation capacity of reinforced concrete subjected to chloride extraction and test the efficiency of re-alkalinization and corrosion inhibitors (Elsener 2001).

It is also worth mentioning the work developed by McCarter and Vennesland (2004) in which the authors applied the open circuit potential as a principle for the development of a corrosion sensor for reinforced concrete structures.

The ultrasonic wave propagation method (UPV) is based on the emission of high frequency waves (above 20 kHz) that propagate through a tested element (Rose 1990). The method is a true non-destructive method, as it does not cause any damage to the analyzed structure. The pulse velocity of the pressure wave (p wave) is applied to the concrete structure by easily generating a wave for easy measurement. It is an extremely promising tool because it allows, among other applications, to evaluate the behavior of concrete in fire situations (Malhotra and Carino 2004; Balaji and Raju 2009), the homogeneity of the concrete, estimate the compressive strength of the concrete, detect the presence of cracks or to determine the dynamic module of the material (Balaji and Raju 2009; Güçlüer 2020; Silva et al. 2021).

However, it should be noted that, even though the UPV test is one of the most popular non-destructive techniques used in the evaluation of concrete properties, there are a number of factors that can affect the experimental values obtained by UPV, namely when measuring the compressive strength of the concrete. Among these factors it is possible we have the different types of cement used, its curing time, the additions made to the cement, the water-cement ratio, the aggregate-cement ratio, etc.

Finally, should be mentioned that the UPV test can be applied through three different forms of transmission: direct, indirect and semi-direct transmission. According to the Brazilian standard NBR 8802 (2019), in direct transmission, waves are received with greater intensity and for this reason it is the most recommended configuration to determine the speed of propagation of waves in a given medium. Indirect transmission is used when there is no access to one of the faces of the element, as the determination of speed through indirect transmission is less satisfactory than through direct transmission.

The main objective of this work is to compare and analyse the experimental results obtained with non-destructive and destructive experimental technics of an extensive campaign performed in thirty concrete columns of a 10 multi-story residential building located in Paraíba, Brazil.

Fig. 1 Case study analysed

2 Experimental Campaign

2.1 *Case Study*

This work presents the experimental results of an extensive non-destructive and destructive in-situ campaign performed in thirty concrete columns of a 10 multi-story residential building (see Fig. 1), located in Paraíba, Brazil, built about 30 years ago, to assess the concrete quality to decide the need of repair or retrofitting works.

The non-destructive and destructive tests were performed to evaluate the depassivation of steel reinforcements and the homogeneity of the concrete, based on: (i) Ultrasound Measurements; (ii) Electrical resistivity of concrete; (iii) Corrosion Potential; (iv) Carbonation Front; (v) Void Index; (vi) Compressive strength and (vii) Capillary absorption.

2.2 *Material and Methods*

The concrete of the building structure exhibited signs of deterioration due to time, notably cracks and detached reinforcement concrete coverings. Taking into account this scenario and considering also the equipment availabilities during in situ inspections performed, it was decided to carry out an Ultrasonic Pulse Velocity test, Electrical resistivity test, and Corrosion potential tests to assess the concrete quality.

2.2.1 Ultrasonic Pulse Velocity

The Ultrasonic Pulse Velocity (UPV) tests were performed on concrete columns according to the Brazilian standard NBR 8802 (2019), using the method of direct transmission between transducers, as described in Fig. 2. For this purpose, the Pundit equipment was used. Lab (manufacturer Proceq), this equipment is composed of two transducers that support the measurement (a wave transmitter and a receiver). Before starting the extensive experimental measurement campaign, the equipment had to be configured for a 54 kHz transmit/receive frequency, through transducers with a diameter of 50 mm.

Before each measurement, the surface to be analysed must be cleaned to remove any material that could mask the penetration of the sonic beam. In addition, a coupling gel must be used in order to improve the contact between the transducers and the surfaces of the concrete to be analysed. Table 1 presents an overall indication of the concrete quality according to the measured ultrasonic pulse velocity.

Fig. 2 Ultrasonic pulse velocity test

Table 1 Quality of concrete as a function of the UPV (Whitehurst 1951)

UPV (m/s)	> 4500	3500–4500	3000–3500	2000–3000	< 2000
Concrete quality	Excellent	Generally good	Questionable	Generally poor	Very poor

Table 2 Corrosion probability of concrete as a function of the electrical resistivity (RILEM TC 2000)

Electrical resistivity (kΩcm)	> 100	50–100	10–50	≤ 10
Corrosion probability	Negligible	Low	Moderate	High

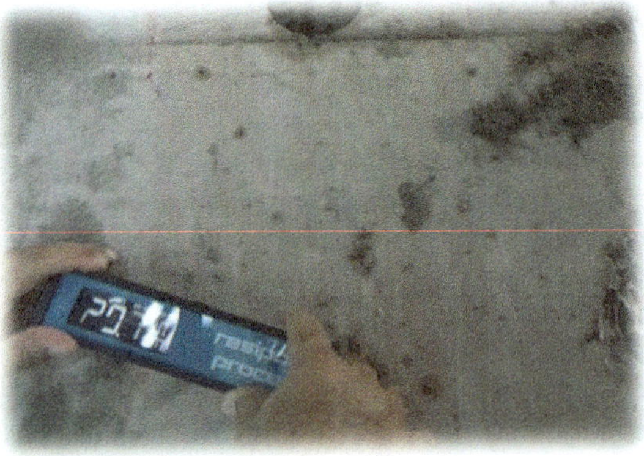

Fig. 3 Electrical resistivity test and (c) corrosion potential test

2.2.2 Electrical Resistivity

Since non-destructive techniques (NDTs) can provide data on the structure without the necessity of sampling the structure (therefore exposing the structure itself to higher risks of damage), they are suitable for Structural Health Monitoring. Among NDTs, the measurement of electrical resistivity of concrete is increasing wider and wider its applicability among the scientific and technical community. In accordance with this method, the surface resistivity measurements were performed with the Proceq Resipod equipment (distance of 50 mm between the electrodes), according to RILEM TC 154 (2000). The equipment was configured with standard parameters (no form factor was included—factor 1.0 was defined).

Table 2 shows the probability of the occurrence of corrosion in the concrete according to its electrical resistivity. The tests performed are shown in an illustrative way in Fig. 3.

2.2.3 Corrosion Potential

Corrosion potential measurement is an experimental technique used to classify the corrosion probability of carbon steel reinforcement used in concrete. This fast and low-cost technique allows the monitoring of reinforced concrete structures over time (Helene 1993a; Medeiros et al. 2017), in order to identify non-passive steel zones that require analysis or repairs (Medeiros et al. 2013a). For this purpose, the international standard described by ASTM C 876 (2015) must be applied. A study developed by Pradhan and Bhattacharjee (2009) showed that this technique is very useful to verify the initiation of corrosion of reinforcements in concrete structures exposed to chloride-contaminated environments.

The corrosion potential of the reinforcement concretes results from the combination of the oxidation of the metal, and the reduction of dissolved oxygen (Romano et al. 2013).

In a work developed by Song and Saraswarhy (2007), the authors showed that the aqueous solution contained in the pore network of the concrete acts as an electrolyte. This fact causes the steel to develop an electrical potential that varies between locations in the reinforced concrete structure due to discontinuities generated by carbonation processes, chloride contamination, variations in porosity and relative humidity, among others.

The equipment used to measure corrosion potential consists of a voltmeter (capable of recording the measurements of potential difference), a reference electrode and a sponge of high conductivity. Darby et al. (1999) showed the voltmeter can be replaced by a digital voltmeter, as long as it has an impedance of at least 20 MΩ.

In resume, open circuit potential measurement is one of the most widely used methods for rebar corrosion assessment on existing reinforced concrete structures. The most commonly applied standard for this method, ASTM C-876 (2015), describes standard test method for corrosion potentials of uncoated reinforcing steel in concrete and evaluation of results. Thus, the corrosion potential was performed the corrosion potential was performed with the help of equipment Profometer, from manufacturer Proceq (see Fig. 4). Table 3 presents the corrosion probability of concrete as a function of the corrosion potential.

Finally, the principal mechanical property (compression strength) and cement qualities (water capillary absorption, void index and density) were performed in all samples analysed in accordance with Brazilian standards NBR 7680-1 (2015) and NBR 9778 (2005), respectively.

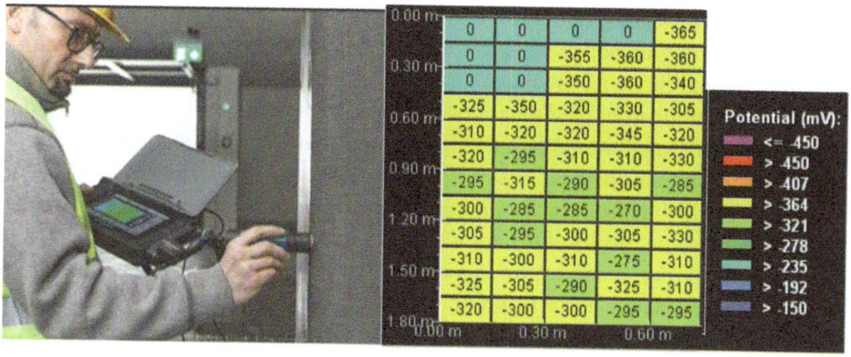

Fig. 4 Corrosion potential test

Table 3 Corrosion probability of concrete as a function of the corrosion potential NBR 7680-1 (2015)

Corrosion potential, Cu/SO_4Cu (mV)	< – 200	– 200 to – 350	> – 350
Corrosion probability	> 10%	Uncertain	> 90%

3 Results and Discussion

Several inspection methods can be used to assess the corrosion state of steel reinforcement in concrete. Especially for periodical field surveys and monitoring, non-destructive testing (NDT) methods are always preferable because they do not generate any or very limited damage to the concrete element investigated. The experimental results are presented in Table 4.

Analyzing the results of the tests performed, it can be seen that for more than 90% of the columns investigated, corrosion potentials were obtained in the range between – 200 mV and 350 mV. This fact characterizes an uncertain probability of corrosion occurrence, according to ASTM C-876 (2015), and it is not possible to state only with the data from this test if there are or not risks of corrosion for the reinforcement of the columns. On the other hand, when the results of the electrical resistivity are observed for these same columns, it can be seen that the values obtained are in the range of 10–50 kΩcm, a situation that characterizes a moderate risk of the manifestation of the corrosion phenomenon. This fact shows that the performance of more than one type of test to verify the corrosion risk is a more efficient strategy for the diagnosis of the quality of a reinforced concrete structure than the use of only one test for this purpose.

The results of the ultrasound tests, although they do not offer direct quantitative information about the risk of reinforcement corrosion, are useful to have an overview of the quality of the concrete and, in this sense, the results show that the concrete of most of the studied columns exhibited a moderate quality, thus validating the results of the resistivity and corrosion potential tests. It was possible to observe a widespread corrosion of the building's support structure, with some columns exhibiting an advanced state of degradation and others with depassivation of reinforcement, meaning the onset of the corrosive process.

The investigation identified the following anomalies: reinforcement corrosion due to depassivation, low-strength concrete and inadequate reinforcement coverage. In view of such irregularities, it is assumed that the concrete structure has a low useful life, thus not reaching the durability parameters provided for in the current standards.

The experimental results of the non-destructive tests presented in Table 4 showing a low quality of the concrete. In accordance with these results, it was necessary to carry out destructive tests to know the mechanical properties of the concrete (compression strength, void index, water absorption and density), as presented in Table 5. The samples selection for the destructive tests was considered where the concrete visually presented a better state of conservation.

Table 4 Experimental results of ultrasound measurements, electrical resistivity and corrosion potential of the reinforcements (average values)

Sample	Electrical resistivity (kΩcm)	Ultrasound (m/s)	Corrosion potential (-mV)
P1	18	2608	355
P2	16	2470	243
P3	12	3450	371
P4	23	3250	266
P5	32	3370	360
P6	15	3300	193
P7	17	3467	243
P8	24	2547	192
P9	19	2470	245
P10	19	1970	243
P11	17	3480	395
P12	24	3380	243
P13	19	2930	240
P14	12	2300	205
P15	24	2930	315
P16	15	2360	285
P17	16	2470	228
P18	19	2870	239
P19	17	1890	252
P20	21	3160	243
P21	17	2300	243
P22	17	3300	243
P23	17	1300	290
P24	13	2630	243
P25	20	2100	229
P26	14	2890	334
P27	21	1300	243
P28	21	3100	243
P29	12	2230	243
P30	16	1700	243

According to Helene (1993b), the concrete samples with water absorption greater than 6.3% and void index greater than (IV) 15% are considered deficient. The studied concrete samples present average results of 17.1% for water absorption and 30.6% for void index (see Table 5). The obtained parameters, above recommendations, showed that these samples of concrete are deficient, corroborating the results found in the non-destructive tests (see Table 5).

Table 5 Experimental results of density, void index, compressive strength and capillarity absorption tests

Sample	Density (g/cm^3)	Capillary absorption (%)	Void index (%)	Compressive strength (MPa)
P15	2.94	14.93	30.5	12.7
P15	2.17	18.74	28.9	12.1
P13	2.70	17.63	32.3	10.9

It is important to remark that the results obtained showed that the pathological manifestations observed in the in situ inspections performed revealed a not-so-good concrete in terms of compressive strength—the average result found in the three concrete core samples extracted from a column located in the garage was 11.9 MPa. At the time the building was constructed (30 years ago), the usual concrete compressive strength specified in concrete designs in the region ranged from 15 to 20 MPa. Taking into account this information and comparing them with the results obtained from the tests performed, one can speculate that either the concrete was not produced with adequate strength at the time or there was no compressive strength evolution over time, or, still, that the pathological manifestations observed contributed to decreasing the initial compressive strength of the concrete.

4 Conclusions

This work discussed the results of an experimental campaign performed on 30 columns of a multifamily reinforced concrete building built about 30 years ago. Non-destructive tests of electrical resistivity, corrosion potential and ultrasonic pulse velocity were performed. The results obtained were efficient to capture the corrosion risk of reinforcement when the electrical resistivity, corrosion potential, and ultrasonic pulse velocity tests are performed together. The analysis of results through this strategy allows the understanding of aspects that are not visible when only one type of test is performed to obtain the reinforcement corrosion risk in wired concrete elements.

Acknowledgements This work was supported by: Base Funding—UIDB/04708/2020 and Programmatic Funding—UIDP/04708/2020 of the CONSTRUCT—Instituto de I&D em Estruturas e Construções—funded by national funds through the FCT/MCTES (PIDDAC) and by FCT—Fundação para a Ciência e a Tecnologia through the individual Scientific Employment Stimulus 2020.00828.CEECIND.

References

Andrade C, Alonso C (2001) On-site measurements of corrosion rate of reinforcements. Constr Build Mater 15(2–3):141–145

Assouli B, Ballivy G, Rivard P (2008) Influence of environmental parameters on application of standard ASTM C876–91: half cell potential measurements. Corros Eng, Sci Technol 43(1):93–96

ASTM C-876 (2015) Standard test method for corrosion potentials of uncoated reinforcing steel in concrete. ASTM International, West Conshohocken, PA

Balaji KVGD, Raju SSSVG (2009) Nondestructive assessment of concrete structures exposed to fire. Int J Appl Environ Sci 4:25–31

Baroghel-Bouny V, Belin P, Maultzsch M, Henry D (2007) $AgNO_3$ spray test: advantages, weaknesses, and various applications to quantify chloride ingress into concrete, Part 2: non-steady-state migration tests and chloride diffusion coefficients. Mater Struct 40:783–799

Broomfield JP, Davies K, Hladky K (2002) The use of permanent corrosion monitoring in new and existing reinforced concrete structures. Cement Concr Compos 24(1):27–34

Castro-Borges P, Ordaz JM (2009) Corrosion performance of concrete columns after localized repairs in a tropical coastal environment. Corrosion 65(2):118–125

Castro-Borges P, Torres-Acosta A, Balancanzapata M, Lopez-Gonzalez A (2012) Análisis de daños por la interacción de cargas ambientales y estructurales en una subestructura marina. Rev Construcción 11(3):88–98

Dantas Neto SA (2004) Evaluation of the properties of binders and asphalt mixtures modified with granulated rubber of used tires. PhD Thesis, University of Brasilia, Brazil

Darby J, Capeling G, George C, Dill M, Hamersley G (1999) Testing and monitoring the durability of concrete structures. Draft—Concrete bridge development group—Technical guide no 2, London, UK

Dinakar P, Sahoo PK, Sriram G (2013) Effect of metakaolin content on the properties of high strength concrete. Int J Concr Struct Mater 7:215–223

El-Din HKS, Eisa AS, Aziz BH, Ibrahim A (2017) Mechanical performance of high strength concrete made from high volume of Metakaolin and hybrid fibers. Constr Build Mater 140:203–209

Elsener B (2001) Half-cell potential mapping to assess repair work on RC structures. Constr Build Mater 15:133–139

Elsener B, Andrade C, Gulikers J, Polder R, Raupach M (2003) Half-cell potential measurements—potential mapping on reinforced concrete structures. Mater Struct 36:461–471

Feliu S, González JA, Miranda JM, Feliu V (2005) Possibilities and problems of in situ techniques for measuring steel corrosion rates in large reinforced concrete structures. Corros Sci 47(1):217–238

Güçlüer K (2020) Investigation of the effects of aggregate textural properties on compressive strength (CS) and ultrasonic pulse velocity (UPV) of concrete. J Build Eng 27:100949

He F, Shi C, Yuan Q, Chen C, Zheng K (2012) $AgNO_3$—based colorimetric methods for measurement of chloride penetration in concrete. Constr Build Mater 26:1–8

Helene PRL (1993a) Contribuição ao estudo da corrosão em armaduras de concreto armado. 231f. Tese (Livre docência)—University of São Paulo (USP), São Paulo

Helene PR (1993b) Contribuição ao estudo da corrosão em armaduras de concreto armado. São Paulo 231:14

Helene P, Pereira F, Castro P (2006) Performance of a 40-year-old concrete bridge with embedded, prestressed galvanized strands. Mater Perform 43(10):42–45

Liam KC, Roy SK, Wood DON (1992) Chloride ingress measurements and corrosion potential mapping study of a 24-year-old reinforced concrete jetty structure in a tropical marine environment. Mag Concr Res 44(160):205–215

Malhotra VM, Carino NJ (2004) Handbook on nondestructive testing of concrete, 2nd edn. CRC Press, Boca Raton, FL, USA

McCarter WJ, Vennesland Ø (2004) Sensor systems for use in reinforced concrete structures. Constr Build Mater 18:351–358

Medeiros MHF, Giordano DE, Pereira E, Vignolio A, Galeano R, Helene P (2012) Inspeção no palácio de la luz - Montevideo: uma visão de durabilidade. Revista Alconpat 2:94–109

Medeiros MHF, Knuutila M, Pereira E, Helene P (2013a) Inspection of buildings in Rio de Janeiro-Brazil: proving the greater tendency of corrosion at the base of reinforced concrete columns using potential corrosion technique. Am J Eng Res (AJER) 2:102–112

Medeiros MHF, Gobbi A, Réus GC, Helene P (2013b) Reinforced concrete in the marine environment: effect of wetting and drying cycles, height and positioning in relation to the seashore. Constr Build Mater 44:452–457

Medeiros MHF, Rocha FC, Medeiros-Junior RA, Helene P (2017) Corrosion potential: influence of moisture, water-cement ratio, chloride content and concrete cover. Rev IBRACON Estrut Mater 10(4):864–885

Medeiros-Junior RA, Lima MG (2016) Electrical resistivity of unsaturated concrete using different types of cement. Constr Build Mater 107:11–16

Metha PK, Monteiro PJM (2014) Concrete: microstructure, properties and materials (4th ed). McGraw-Hill Education, New York

NBR 9778 (2005) Hardened mortar and concrete—determination of absorption, voids and specific gravity. ABNT—Associação Brasileira de Normas Técnicas

NBR 7680-1 (2015) Concrete—sampling, preparing, testing and result analysis of concrete cores part 1: axial compressive strength. ABNT—Associação Brasileira de Normas Técnicas

NBR 8802 (2019) Hardened concrete—determination of ultrasonic wave transmission velocity. ABNT—Associação Brasileira de Normas Técnicas

Poupard O, L'hostis V, Catinaud S, Petre-Lazar I (2006) Corrosion damage diagnosis of a reinforced concrete beam after 40 years natural exposure in marine environment. Cem Concr Compos 36(3):504–520

Pradhan B, Bhattacharjee B (2009) Half-cell potential as an indicator of chloride-induced rebar corrosion initiation in RC. J Mater Civ Eng 21(10):543–552

Ray I, Gong Z, Davalos JF, Kar A (2012) Shrinkage and cracking studies of high performance concrete for bridge decks. Constr Build Mater 28:244–254

RILEM TC 154 (2000) Electrochemical techniques for measuring metallic corrosion test methods for on site measurement of resistivity of concrete. Mater Struct 33:603–611

Romano P, Brito PSD, Rodrigues L (2013) Monitoring of the degradation of concrete structures in environments containing chloride ions. Constr Build Mater 47:827–832

Rose JL (1990) Ultrasonic waves in solid media. Cambridge University Press, Cambridge, UK

Setzer MJ, Heine P, Kasparek S, Palecki S, Auberg R, Feldrappe V, Siebel E (2004) Test methods of frost resistance of concrete: CIF-test: capillary suction, internal damage and freeze thaw test - Reference method and alternative methods A and B. Mater Struct 37:743–753

Silva FAN, Delgado JMPQ, Cavalcanti RS, Azevedo AC, Guimarães AS, Lima AGB (2021) Use of nondestructive testing of ultrasound and artificial neural networks to estimate compressive strength of concrete. Buildings 11(2):44

Song H, Saraswarhy V (2007) Corrosion monitoring of reinforced concrete structures—a review. Int J Electrochem Sci 2:1–28

Tang L (1996) Chloride transport in concrete—measurement and prediction. PhD Thesis, Chalmers University of Technology, Gothenburg, Sweden

Whitehurst E (1951) Soniscope tests concrete structures. J Am Concr Inst 47:433–444

Zhang T, Gjørv OE (1994) An electrochemical method for accelerated testing of chloride diffusivity in concrete. Cement Concrete Res 24:1534–1548

Electrical Resistivity and Carbonation Front of LC^3 Concretes Incorporating Different Supplementary Cementitious Materials

C. E. T. Balestra, G. Savaris, R. Schneider, A. Y. Nakano, and M. H. Pietrobelli

Abstract Due to the clinkerization process during the Portland cement production, large amounts of CO_2 are emitted, increasing the effects related to climate change, consequently, the seek for alternatives to mitigate these emissions are necessary. The use of supplementary cementitious materials (SCM) to partially replace Portland clinker/cement has been the subject of different research, including the use of LC^3 (Limestone Calcined Clay Cements), where up to 50% of Portland clinker can be replaced. However, the cement industry has already used other SCM with pozzolanic activities in commercial cement and the interaction in LC^3 still needs contributions. In this sense, this work evaluates the performance of concretes containing LC^3 mixtures incorporating different SCM (silica fume, fly ash, sugarcane bagasse ash, and acai stone ash) regarding its durability by volumetric electrical resistivity and accelerated carbonation. The results showed that the presence of SCM in LC^3 concretes increases the resistivity to ionic flow probably due to a refinement in the concrete microstructure, whereas, for carbonation, all concrete with LC^3 presented higher carbonation fronts in relation to the reference concrete due to the low Portlandite availability to react with the CO_2 that penetrates into the concrete pores.

C. E. T. Balestra (✉) · G. Savaris
Department of Civil Engineering, Federal University of Technology—Parana, Curitiba, Brazil
e-mail: carlosbalestra@utfpr.edu.br

G. Savaris
e-mail: gsavaris@utfpr.edu.br

C. E. T. Balestra
Department of Environmental Sciences, Western Parana State University, Cascavel, Brazil

R. Schneider
Department of Chemistry, Federal University of Technology—Parana, Curitiba, Brazil
e-mail: rschneider@utfpr.edu.br

A. Y. Nakano · M. H. Pietrobelli
Department of Electronic Engineering, Federal University of Technology—Parana, Curitiba, Brazil
e-mail: nakano@utfpr.edu.br

J. M. P. Q. Delgado (ed.), *Building Pathologies: Experimental Campaigns and Numerical Procedures*, Building Pathology and Rehabilitation 25,
https://doi.org/10.1007/978-3-031-17061-4_6

Keywords LC^3 · Carbonation front · Electrical resistivity · Concrete · CO_2 · Clinker · Supplementary cementitious materials

1 Introduction

In recent years, several researches have been carried out seeking for alternatives to reduce the impacts related to climate change and global warming, being carbon dioxide (CO_2) one of the leading gases responsible for the greenhouse effect. In this context, the Paris Agreement, signed by 195 nations and European Union, aims to limit the rise in temperature up to 1.5 °C in relation to pre-industrial levels. At this point, the clinkerization process in the cement industries is responsible for approximately 5–10% of global CO_2 emissions, where for each ton of cement produced, 850–1100 kg of CO_2 are emitted. From this perspective, it is estimated that cement consumption will increase, not only for new infrastructure, but also due to the necessity for conservation and rehabilitation of existing structures, which aggravates the problem related to the emission of CO_2 from cement production that contributes to environmental impacts (Mehta and Monteiro 2006; Scrivener et al. 2018; Nair et al. 2020; Krishnan and Bishnoi 2020; Yu et al. 2021; Antoni et al. 2012; Sánches Berriel et al. 2016).

One of the alternatives to mitigate CO_2 emissions is related to the use of supplementary cementitious materials (SCM) to partially replace Portland clinker in cement. Nair et al. (2020) pointed out a variation between 0.81 and 0.64 kg CO_2/kg of cement when SCM is used. In this sense, blast furnace slag and fly ash, which are residues from the steel and energy industries are the primary SCM used (Du and Pang 2020; Scrivener et al. 2019; Avet et al. 2016; Zhang et al. 2020). Blast furnace slag has latent hydraulic properties that are, present in an alkaline environment with water. The hydration reactions lead to the formation of hydration products, analogous to that observed in the hydration of Portland clinker, providing strength and durability to the concrete. Fly ash has pozzolanic properties, leading to the formation of C–S–H from reactions with portlandite ($Ca(OH)_2$ or also represented by CH) (Mehta and Monteiro 2006; Neville 2020). Although the effects of both (blast furnace slag and fly ash) are recognized in the literature, the availability to meet the rising demand for Portland cement is insufficient globally, since all blast furnace slag and fly ash with pozzolanic properties are already been used by cement industries (Scrivener et al. 2018, 2019; Nair et al. 2020; Du and Pang 2020).

In this context, the search for alternatives to the current cement production has motivated research regarding the combined use of metakaolin and limestone for partial replacement of Portland clinker. Limestone Calcined Clay Cements (LC^3-50) reach replacements up to 50% of Portland clinker are reported in ternary mixtures of metakaolin (Kaolinitic calcined clay) and limestone. Basically, LC^3-50 cements are composed of 50% Portland Clinker, 30% Metakaolin, 15% Limestone and 5% Gypsum, allowing to reduce significantly the amount of Portland Clinker present in the cement (Scrivener et al. 2018, 2019; Nair et al. 2020; Krishnan and Bishnoi 2020;

Yu et al. 2021; Antoni et al. 2012; Sánches Berriel et al. 2016; Du and Pang 2020; Avet et al. 2016; Zhang et al. 2020). Furthermore, the clay calcination temperature is between 650 and 800 °C (half the temperature to produce Portland clinker) and does not involve limestone decarbonation, mitigating CO_2 emissions in the cement industry (Krishnan and Bishnoi 2020).

Indeed, Sánches Berriel et al. (2016) pointed out that even applying wet production processes for the LC^3-50, the performance against global warming surpasses the most modern cement production processes today. Almenares et al. (2017) analyzed variations in the production of calcined clays using the infrastructure of cement industries, looking for the production of LC^3. The authors successfully demonstrated the feasibility of producing calcined clay with suitable properties for the production of LC^3 using the existing infrastructure of cement industries. Krishnan and Bishnoi (2020) pointed out that clays with kaolinite content of 27% are already suitable for the production of LC^3. Scrivener et al. (2019) emphasize that clays with kaolinite content between 40 and 75% lead to greater significance in mechanical strength. Consequently, it is unnecessary the use calcined clay with kaolinitic content above this range. Avet et al. (2016) presented a study where, through thermogravimetry assays, the pozzolanicity of calcined clay can be evaluated through the loss of water between 110 and 400 °C.

In LC^3-50 mixtures, the aluminate phase present in the calcined clay consumes Portlandite (CH) forming hemicarbonate (metastable phase) and mono-carboaluminates (final stable phase) that control limestone reaction. From this mechanism, a refinement of the cement matrix, enhancing the mechanical strength and durability of concrete is obtained (Yu et al. 2021; Avet et al. 2016). Rodriguez and Tobon (2020) pointed out that clinker hydration process in the LC^3 has an influence on the compressive strength at all ages. On the other hand, the calcined clay/limestone ratio has influence only 72 h after the beginning of hydration due to the intense pozzolanic activity.

Although research with LC^3 has advanced in recent years, there are still lines of research that need contributions, such as the incorporation of different SCM such as silica fume, fly ash, sugarcane bagasse ash, or acai stone ash in mixtures containing metakaolin and limestone. At this point, not only the mechanical performance but also issues related to durability must be evaluated, reducing the costs related to maintenance and rehabilitation and the consumption of raw materials.

In this sense, concrete electrical resistivity and carbonation are important parameters analyzed in order to evaluate the service life of reinforced concrete structures from the perspective of reinforcement corrosion risk. Reinforcement corrosion is an electrochemical process and one of the main issues associated with field reinforced concrete structures deterioration, whether in an urban or marine environment, reducing the bearing capacity of the structures, due to a decrease in mechanical properties of reinforcement as a function of the increase in the corrosion degree. In this sense, chlorides (marine environment) and/or CO_2 (urban environment) penetrate through concrete pores and, when achieving reinforcement surface, end up destroying the passivating layer that protects reinforcements in concrete and provides the conditions for the beginning of the corrosive process (Balestra et al. 2016; Rehman and Al-Hadhrami 2013; Khan et al. 2014; Apostolopoulos et al. 2013).

Electrical resistivity is a material property that describes its ability to handle the charge flow through the gradient between apparent voltage and current passing, both multiplied by a constant that describes the material's properties (Yu et al. 2017; Chen et al. 2014; Sengul 2014). Considering that reinforcement corrosion is an electrochemical process, the electrical resistivity is a non-destructive technique that allows continuous monitoring of field concrete structures in order to evaluate the probable reinforcement corrosion risk. Horbostel et al. (2013) presented an extensive literature review regarding the relationship between electrical resistivity and corrosion risk. Sengul (2014) describes electrical resistivity as an indicator of the durability of concrete structures. Bem et al. (2018) describe parameters relevant to concrete characteristics (aggregates, cement consumption, porosity, presence of SCM) and their effects on electrical resistivity. Medeiros-Junior and Lima (2016), Hou et al. (2017). Balestra et al. (2019) correlated the electrical resistivity of field reinforced concrete structures with the penetration of chlorides, potentially causing reinforcement corrosion.

Carbonation deals with the concrete pH reduction by the CO_2 action that ends up destroying the passivating film that protects the reinforcements in concrete, creating conditions to start the corrosive process. At this point, some of the main lines of research presented in the literature range from concrete dosage and use of SCM (Bucher et al. 2017; Shi et al. 2016; Silva et al. 2015) to numerical modeling of CO_2 penetration (Paul et al. 2018). Regarding the analysis of field concrete structures, Han et al. (2013) analyzed the carbonation of 20-year-old coastal structures, with carbonation fronts between 3 and 15 mm being observed. Talukdar and Banthia (2013) analyzed the carbonation front of buildings in different cities around the world in the 2000s, with carbonation fronts between 3 and 13 mm.

Although reinforcement corrosion is well-documented in the literature, there is a gap regarding the analysis of durability considering the use of concrete with lower carbon emission cement, such as LC^3-50 incorporating different SCM. Thus, the present work fits into this perspective, seeking to analyze how the presence of SCM (Silica fume, fly ash, sugarcane bagasse ash, and acai ash) can affect both the electrical resistivity and the carbonation front of LC^3 concretes in order to evaluate the reinforcement corrosion risk as described in next sections.

2 Materials and Methods

The analysis of different types of SCM (silica fume, fly ash, sugarcane bagasse ash, and acai stone ash) in the electrical resistivity and carbonation front LC^3 concrete was performed in this work. Therefore, the reference mix was 1: 2.35: 2.85: 0.63 (cement: fine aggregate: coarse aggregate: water) with slump 100 $\pm$ 20 mm with approximately 350 kg/m^3 of cement content. The aggregate properties are shown

Table 1 Aggregate's characteristics

Identification	Specific mass (kg/m^3)	Maximum characteristic dimension (mm)	Fineness modulus
Coarse aggregate	2680	9.50	–
Fine aggregate	2650	0.60	1.95

in Table 1. Given the difficulty in obtaining clinker to perform the tests, a Brazilian Portland cement CP V ARI type (Analogous to ASTM TYPE III or CEM I) was used in this study. In this sense, clinker, gypsum and limestone content of this cement was informed by the manufactory, and corrections for LC^3 obtention were performed. The use of Portland cement to produce LC^3 mixtures was also performed in different works presented in the literature (Yu et al. 2021; Du and Pang 2020; Zhang et al. 2020).

In addition, it is worth mentioning that to achieve a slump greater than 80 mm in LC^3 mixtures, a dosage equal to 3.5% of polycarboxylate ether-based superplasticizer (PCE) in relation to the binder mass was required. The literature (Nair et al. 2020; Zaribaf et al. 2015; Ferreiro et al. 2017) had already pointed out the necessity of superplasticizers in LC^3 mixtures. Finally, due to the presence of metakaolin and limestone in the mixtures with LC^3, the SCM content was standardized at 10% making up the percentage relationships shown in Table 2. The gypsum content was standardized at 5% for all mixtures. The chemical characteristics of the materials determined by Energy Dispersive Spectroscopy (EDS) are shown in Table 3.

After this mixing, 10 × 20 cm cylindrical specimens were molded and cured under saturated conditions at 23 ± 2 °C for 28 days. After this curing period, six specimens of each mixture were conducted for the volumetric electrical resistivity measurement setup (Fig. 1a) and for the accelerated carbonation assay (Fig. 1b). For electrical resistivity, concretes were washed in running water in order to remove calcium present on their surface after the curing procedure and were subsequently subjected to the electrical resistivity test as recommended by ASTM C 1876 (2019). For carbonation, the maximum CO_2 penetration occurs when the pore saturation

Table 2 Group's identification and perceptual of SCM

Identification	Portland cement (%)	Calcined clay (%)	Limestone (%)	SCM (%)			
				Silica fume	Fly ash	Sugarcane bagasse ash	Acai stone ash
Reference	100	–	–	–	–	–	–
LC^3-50	55	30	15	–	–	–	–
LC^3-60 S	45	30	15	10	–	–	–
LC^3-60 FA	45	30	15	–	10	–	–
LC^3-60 SBA	45	30	15	–	–	10	–
LC^3-60 ASA	45	30	15	–	–	–	10

Table 3 Main chemical elements of the cement and SCM

Material	Si	Ca	C	S	Mg	K	Na	Al	Fe
Cement	8.58	25.35	5.45	1.83	3.55	1.39	0.15	3.23	1.42
Calcined clay	9.65	–	15.24	–	1.15	0.35	–	33.53	1.82
Limestone	1.25	24.11	14.85	–	2.19	0.26	–	–	0.19
Silica fume	44.58	–	–	–	0.64	0.63	–	–	–
Fly ash	18.87	0.85	20.57	0.71	1.58	1.90	0.57	–	1.40
Sugarcane bagasse ash	11.34	1.83	33.01	–	2.04	2.67	–	–	7.51
Acai stone ash	2.07	2.08	50.46	3.30	2.35	9.43	–	–	0.13

degree is close to 50% (Comité Euro-International du Béton 1992). In this way, to reach this saturation degree, the specimens were placed in an oven at 80 °C, with the mass being determined at intervals of 60 min up to reach the saturation degree of 50%. After the concrete specimens were placed in the carbonation chamber and were maintained for 48 h at 20Psi CO_2 pressure, being split after this period and sprayed with a phenolphthalein solution as recommended by RILEM CPC 18 (1988). A metal ruler was used for measuring, the carbonation front (Colorless front).

A group of two specimens of each mixture was analyzed in order to verify their concrete alkalinity. In this case, specimens with a 50% saturation degree were not subjected to accelerated carbonation assay, being split and subjected to spraying of a phenolphthalein solution. The alkalinity was verified through the intense pink color observed immediately after spraying the solution, as exemplified in Fig. 2. All concrete (Reference, LC^3-50 and LC^3-60) showed a high alkalinity environment for reinforcement protection.

Micrographs and calorimetry assays were performed in order to support the results. In this sense, cubic one mm side samples (for micrographs) and powder

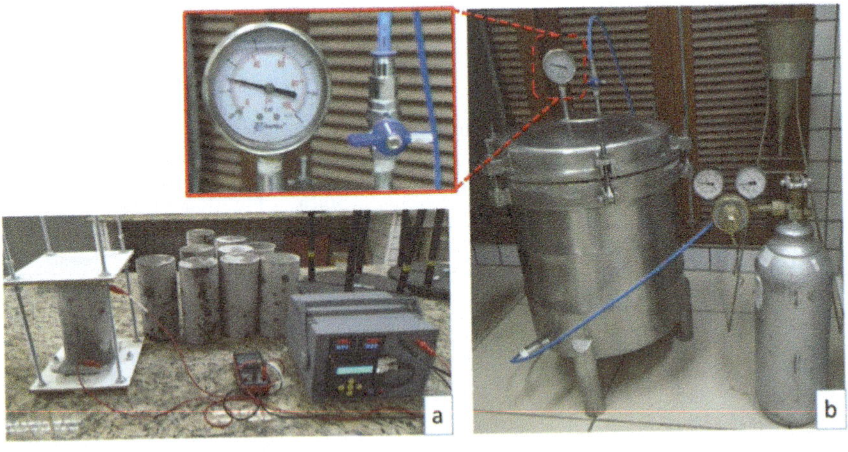

Fig. 1 **a** Electrical resistivity test. **b** Accelerated carbonation assay

Fig. 2 Concrete alkalinity verification: reference (left), LC³-50 (center), and LC³-60 S (right)

samples (for calorimetry assays) from uncarbonated concretes were obtained. The cubic samples were obtained by cutting the concretes with a saw, whereas, powder samples were obtained with a rotary drill. The powder samples were passed in a sieve (#0.075 mm) and were analyzed in a Toledo Mettler Calorimetry from 30° to 900 °C with a range of 10 °C/min in a nitrogen atmosphere (50 ml/min).

3 Results and Discussion

Table 4 shows the result of electrical resistivity, whereas Fig. 3 shows the specimens after spraying the phenolphthalein solution for carbonation front evaluation. It should be pointed in Fig. 3 that when the pink color was not observed in some LC³-60 mixtures a fractured concrete with Ordinary Portland Cement (OPC) was sprayed with the phenolphthalein solution only for color comparison.

The electrical resistivity results show that all concretes presented resistivity values higher than 100 kΩ cm, which, according to the RILEM TC 154-EMC (2010), represents a negligible probability of reinforcement corrosion risk. At this point, the reference concrete had the highest electrical resistivity (348 kΩ cm), while LC³ concretes presented resistivity values between 218 and 324 kΩ cm, demonstrating that concretes with LC³ can be interpreted as durable concrete from the perspective of electrical resistivity, despite having lower values than the reference concrete. It is also observed that in LC³ concretes incorporating SCM (LC³-60 group) the

Table 4 Electrical resistivity results

Identification	Electrical resistivity (kΩ.cm)
Reference	348.92
LC³-50	218.48
LC³-60 S	324.07
LC³-60 CV	268.59
LC³-60 C	247.11
LC³-60A	248.55

Fig. 3 Concrete carbonation front analysis

electrical resistivity increased, even with a smaller amount of Portland clinker in the mixture, demonstrating that the presence of SCM is beneficial by increasing the resistivity to the ionic flux through concrete pores. On the other hand, the carbonation fronts observed after the accelerated assays (Fig. 3) were, in all cases, higher than the carbonation fronts observed in the reference concrete. In this case, all LC^3-50 and LC^3-60 concretes were carbonated (pH < 9) where the colorless was noticed after spraying the phenolphthalein solution, demonstrating a lower performance than reference concrete.

Such behavior may, at first, lead to a contradiction in relation to the durability of concretes containing LC^3. From the perspective of electrical resistivity, LC^3 concretes presents values that refer to a negligible reinforcement corrosion risk while from the perspective of carbonation, the LC^3 concretes had a lower performance than reference concrete, with high carbonation fronts. However, the porosity and the CH presence in concretes with LC^3 must be considered to discuss these observations. In this context, Dhandapani and Santhanam (2017) analyzed the porosity of pastes with OPC and LC^3. LC^3-50 systems have greater resistance to ionic flux due to a reduction in permeability thanks to the refinement of the microstructure. The presence of pozzolanic materials (metakaolin and others SCM) in the LC^3 systems supports a densified matrix, justifying the electrical resistivity values. In fact, the densified matrix of LC^3 mixtures (including the presence of SCM) can be observed in Fig. 4 comparing the micrographs of reference concrete, LC^3-50 and LC^3-60 S where a dense microstructure can be observed.

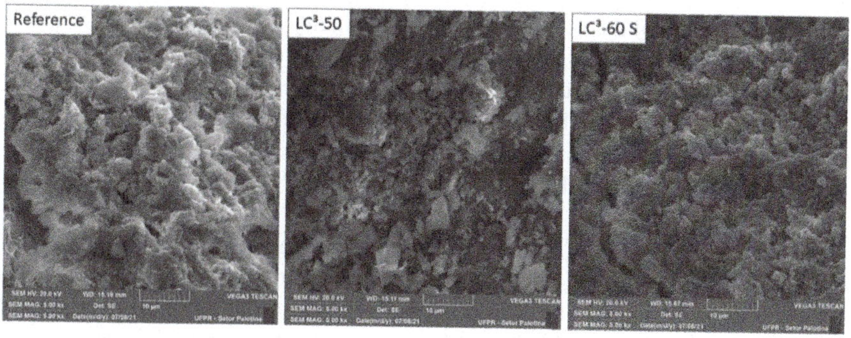

Fig. 4 Micrographs obtained from concretes: reference (left), LC^3-50 (central), LC^3-60 S (right)

Regarding carbonation, other studies had already pointed out higher carbonation fronts in concrete using LC^3 (Scrivener et al. 2018, 2019). In this work, LC^3 systems allow high CO_2 penetration due to the low availability of calcium hydroxide react with CO_2, since it is consumed by the pozzolanic reactions with metakaolin and SCM. Thus, greater carbonation fronts are observed (the presence of $Ca(OH)_2$ is noted in Fig. 5 and will be analyzed below). This observation extends to the LC^3-60 concretes in this work. Although the presence of SCM leads to a refinement of the microstructure and helps to increase electrical resistivity, the adoption of these materials increases the consumption of CH by pozzolanic reactions, further reducing the amount of calcium available to capture the CO_2 that penetrates into the concrete. In this sense, analyzing the different SCM, it is possible to notice in Fig. 3 that only silica fume shows lower carbonation front in relation to the other mixtures, however, still insufficient to provide protection to the reinforcement.

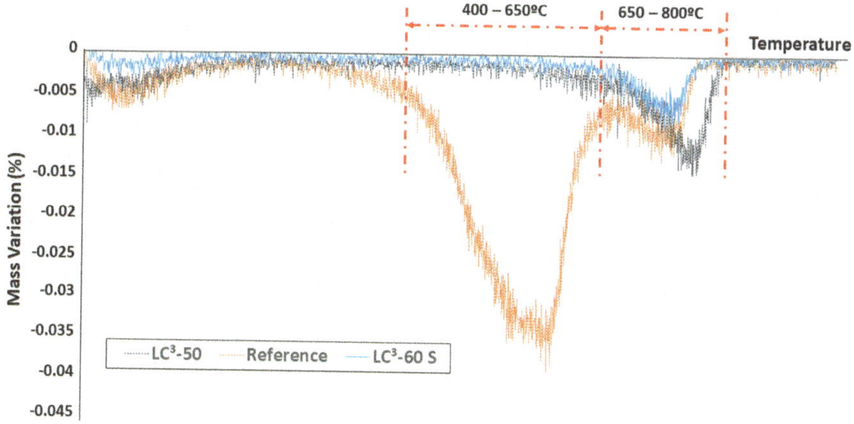

Fig. 5 Calorimetry results of reference (orange), LC^3-50 (gray) and LC^3-60 S (blue) concretes

In order to confirm the statement below, calorimetry analyzes were performed in samples from the specimens uncarbonated. The calorimetry analysis (Fig. 5) shown the absence of the peak associated to CH in LC^3 mixtures (between 400 and 650 °C) confirming that the availability of calcium hydroxide to react with CO_2 that penetrates through concrete pores is limited in LC^3 systems, justifying the greater electrical resistivity and the carbonation fronts observed (Scrivener et al. 2018, 2019; Dhandapani and Santhanam 2017).

Finally, it is possible to affirm that concretes containing LC^3 present high alkalinity and resistivity to ionic flux, where SCM has a beneficial effect on these parameters However, higher carbonation fronts are observed due to the limited availability of calcium hydroxide to react with the CO_2, consequently, the adoption of curing procedures and coating layers are recommended to mitigate CO_2 penetration in order to provide durability to field concrete structures.

4 Conclusions

This work analyzed the effects of the incorporation of different supplementary cementitious materials (SCM—silica fume, fly ash, sugarcane bagasse ash and acai core ash) on the electrical resistivity and carbonation front of LC^3 concretes. The main conclusions are:

- All analyzed concrete (reference, LC^3-50 and LC^3-60 with different SCM) showed high alkalinity, demonstrating that mixtures containing metakaolin, and limestone with or without SCM replacing Portland clinker have the ability to protect reinforcements against corrosion.
- Regarding electrical resistivity, all concretes showed negligible reinforcement corrosion risk with resistivity values above 100 kΩ cm. The SCM contributed to increasing the electrical resistivity by at least 25% compared to the LC^3-50 mixtures. The performance of these concretes containing LC^3 and SCM is due to an improvement in the microstructure in relation to the mixtures containing only metakaolin and limestone.
- The LC^3-50 and LC^3-60 concretes presented higher carbonation fronts in relation to the reference concrete, due to the absence of calcium hydroxide to interact with the CO_2 that penetrates verified by calorimetry assays and in agreement with the literature. Only the mixture containing silica had a lower carbonation front but was still not able to protect the reinforcements.

References

Almenares RS et al (2017) Industrial calcination of kaolinitic clays to make reactive pozzolans. Case Stud Constr Mater 6:225–232. https://doi.org/10.1016/j.cscm.2017.03.005

American Society for Testing and Materials: standard test method for acid-soluble chloride in mortar and concrete. ASTM C1876, Annual Book of ASTM Standards, West Conshohocken (2019)

Antoni M et al (2012) Cement substitution by a combination of metakaolin and limestone. Cem Concr Res 42:1579–1589. https://doi.org/10.1016/j.cemconres.2012.09.006

Apostolopoulos CA, Demis S, Papadakis VG (2013) Chloride-induced corrosion of steel reinforcement—mechanical performance and pit depth analysis. Constr Build Mater 38:139–146. https://doi.org/10.1016/j.conbuildmat.2012.07.087

Avet F et al (2016) Development of a new rapid, relevant and reliable (R^3) test method to evaluate the pozzolanic reactivity of calcined kaolinitic clays. Cem Concr Res 85:1–11. https://doi.org/10.1016/j.cemconres.2016.02.015

Balestra CET, Lima MG, Silva AR, Medeiros-Junior RA (2016) Corrosion degree effect on nominal and effective strengths of naturally corroded reinforcement. J Mater Civ Eng 28(10):04016103. https://doi.org/10.1061/(ASCE)MT.1943-5533.0001599

Balestra CET, Nakano AY, Savaris G, Medeiros-Junior RA (2019) Reinforcement corrosion risk of marine concrete structures evaluated through electrical resistivity: proposal of parameters based on field structures. Ocean Eng 187:106167. https://doi.org/10.1016/j.oceaneng.2019.106167

Bem DH, Lima DPB, Medeiros-Junior RA (2018) Effect of chemical admixtures on concrete´s electrical resistivity. Int J Build Pathol Adapt 36:174–187. https://doi.org/10.1108/IJBPA-11-2017-0058

Bucher R, Diederich P, Escadeillas G, Cyr M (2017) Service life of metakaolin-based concrete exposed to carbonation—comparison with blended cement containing fly ash, blast furnance slag and limestone filler. Cem Concr Res 99:18–29. https://doi.org/10.1016/j.cemconres.2017.04.013

Chen CT, Chang JJ, Yeih W (2014) The effects of specimen parameters on the resistivity of concrete. Constr Build Mater 71:35–43. https://doi.org/10.1016/j.conbuildmat.2014.08.009

Comité Euro-International du Béton (1992) Bulletin D' information N° 183: durable concrete structures, 2nd edn. Telford Services, Lausanne, SW, 122p

Dhandapani Y, Santhanam M (2017) Assessment of pore structure evolution in the limestone calcined clay cementitious system and its implications for performance. Cement Concr Compos 84:36–47. https://doi.org/10.1016/j.cemconcomp.2017.08.012

Du H, Pang SD (2020) High-performance concrete incorporating calcined kaolin clay and limestone as cement substitute. Constr Build Mater 264:120152. https://doi.org/10.1016/j.conbuildmat.2020.120152

Ferreiro S, Herfort D, Damtoft JS (2017) Effect of raw clay type, fineness, water-to-cement ratio and fly ash addition on workability and strength performance of calcined clay—limestone Portland cements. Cem Concr Res 101:1–12. https://doi.org/10.1016/j.cemconres.2017.08.003

Han SH, Park WS, Yang EI (2013) Evaluation of concrete durability due to carbonation in harbor concrete structures. Constr Build Mater 48:1045–1049. https://doi.org/10.1016/j.conbuildmat.2013.07.057

Hornbostel K, Larsen CK, Geiker MR (2013) Relationship between concrete resistivity and corrosion rate—a literature review. Cem Concr Compos 39:60–72. https://doi.org/10.1016/j.cemconcomp.2013.03.019

Hou TC, Nguyen VK, Su YM, Chen YR, Chen PJ (2017) Effects of coarse aggregates on the electrical resistivity of Portland cement concrete. Constr Build Mater 133:397–408. https://doi.org/10.1016/j.conbuildmat.2016.12.044

Khan I, François R, Castel A (2014) Prediction of reinforcement corrosion using corrosion induced cracks width in corroded reinforced concrete beams. Cem Concr Res 56:84–96. https://doi.org/10.1016/j.cemconres.2013.11.006

Krishnan S, Bishnoi S (2020) A numerical approach for designing composite cements with calcined clay and limestone. Cem Concr Res 138:106232. https://doi.org/10.1016/j.cemconres.2020.106232

Medeiros-Junior RA, Lima MG (2016) Electrical resistivity of unsaturated concrete using different types of cement. Constr Build Mater 107:11–16. https://doi.org/10.1016/j.conbuildmat.2015.12.168

Mehta PK, Monteiro PJM (2006) Concrete: microstructure, properties and materials, 3rd edn. Mc-Graw Hill, New York. https://doi.org/10.1036/0071462899

Nair N et al (2020) A study on fresh properties of limestone calcined clay blended cementitious systems. Constr Build Mater 254:119326. https://doi.org/10.1016/j.conbuildmat.2020.119326

Neville AM (2020) Properties of concrete, 5th edn. Prentice Hall, New York. ISBN-10: 0273755803

Paul SC, Panda B, Huang Y, Garg A, Peng X (2018) An empirical model design for evaluation and estimation of carbonation depth in concrete. Measurement 124:205–210. https://doi.org/10.1016/j.measurement.2018.04.033

Rehman S, Al-Hadhrami LM (2013) Web-based national corrosion cost inventory system for Saudi Arabia. Anti Corros Methods Mater 61(2):77–92. https://doi.org/10.1108/ACMM-04-2013-1254

RILEM Recommendation. TC154-EMC (2010) Electrochemical techniques for measuring metallic corrosion. Mater Struct 33:603–611

RILEM Recommendations. CPC-18 (1988) Measurement of hardened concrete carbonation depth. Mater Struct 21:453–455. ISSN: 1359-5997

Rodriguez C, Tobon JI (2020) Influence of calcined clay/limestone, sulfate and clinker proportions on cement performance. Constr Build Mater 251:119050. https://doi.org/10.1016/j.conbuildmat.2020.119050

Sánches Berriel S et al (2016) Assessing the environmental and economic potential of limestone calcined clay cement in Cuba. J Cleaner Prod 124:361–369. https://doi.org/10.1016/j.jclepro.2016.02.125

Scrivener K et al (2018) Calcined clay limestone cements (LC^3). Cem Concr Res 114:49–56. https://doi.org/10.1016/j.cemconres.2017.08.017

Scrivener K et al (2019) Impacting factors and properties of limestone calcined clay cements (LC^3). Green Mater 7(1):3–14. https://doi.org/10.1680/jgrma.18.00029

Sengul O (2014) Use of electrical resistivity as an indicator for durability. Constr Build Mater 73:434–441. https://doi.org/10.1016/j.conbuildmat.2014.09.077

Shi C, He T, Zhang G, Wang X, Hu Y (2016) Effects of superplasticizers on carbonation resistance of concrete. Constr Build Mater 108:48–55. https://doi.org/10.1016/j.conbuildmat.2016.01.037

Silva RA, Neves R, de Brito J, Dhir RK (2015) Carbonation behavior of recycled aggregate concrete. Cem Concr Compos 62:22–32. https://doi.org/10.1016/j.cemconcomp.2015.04.017

Talukdar S, Banthia N (2013) Carbonation in concrete infrastructure in the context of global climate change: development of a service lifespan model. Constr Build Mater 40:775–782. https://doi.org/10.1016/j.conbuildmat.2012.11.026

Yu B, Liu J, Chen Z (2017) Probabilistic evaluation method for corrosion risk of steel reinforcement based on concrete resistivity. Constr Build Mater 138:101–113. https://doi.org/10.1016/j.conbuildmat.2017.01.100

Yu J et al (2021) Compressive strength and environmental impact of sustainable blended cement with high-dosage limestone and calcined clay (LC2). J Cleaner Prod 278:123616. https://doi.org/10.1016/j.jclepro.2020.123616

Zaribaf BH, Uzal B, Kurtis K (2015) Compatibility of superplasticizers with limestone-metakaolin blended cementitious system. RILEM Bookseries 10, Springer, Calcined Clays for Sustainable Concrete, pp 427–434. https://doi.org/10.1007/978-94-017-9939-3_53

Zhang D et al (2020) Engineered cementitious composites (ECC) with limestone calcined clay cement (LC^3). Cem Concr Compos 114:103766. https://doi.org/10.1016/j.cemconcomp.2020.103766

SPRINGER NATURE

GPSR Compliance

The European Union's (EU) General Product Safety Regulation (GPSR) is a set of rules that requires consumer products to be safe and our obligations to ensure this.

If you have any concerns about our products, you can contact us on ProductSafety@springernature.com

In case Publisher is established outside the EU, the EU authorized representative is:

Springer Nature Customer Service Center GmbH
Europaplatz 3
69115 Heidelberg, Germany